WORLD BANK TECHNICAL PAPER NUMBER 55

# Techniques for Assessing Industrial Hazards

## A Manual

Technica, Ltd.

The World Bank
Washington, D.C., U.S.A.

Technical Papers are not formal publications of the World Bank, and are circulated to encourage discussion and comment and to communicate the results of the Bank's work quickly to the development community; citation and the use of these papers should take account of their provisional character. The findings, interpretations, and conclusions expressed in this paper are entirely those of the author(s) and should not be attributed in any manner to the World Bank, to its affiliated organizations, or to members of its Board of Executive Directors or the countries they represent. Any maps that accompany the text have been prepared solely for the convenience of readers; the designations and presentation of material in them do not imply the expression of any opinion whatsoever on the part of the World Bank, its affiliates, or its Board or member countries concerning the legal status of any country, territory, city, or area or of the authorities thereof or concerning the delimitation of its boundaries or its national affiliation.

Because of the informality and to present the results of research with the least possible delay, the typescript has not been prepared in accordance with the procedures appropriate to formal printed texts, and the World Bank accepts no responsibility for errors.

The most recent World Bank publications are described in the catalog *New Publications*, a new edition of which is issued in the spring and fall of each year. The complete backlist of publications is shown in the annual *Index of Publications*, which contains an alphabetical title list and indexes of subjects, authors, and countries and regions; it is of value principally to libraries and institutional purchasers. The latest edition of each of these is available free of charge from the Publications Sales Unit, Department F, The World Bank, 1818 H Street, N.W., Washington, D.C. 20433, U.S.A., or from Publications, The World Bank, 66, avenue d'Iéna, 75116 Paris, France.

Technica, Ltd., London specializes in hazard and risk assessment and accident prevention in petrochemical and chemical installations.

**Library of Congress Cataloging-in-Publication Data**

Techniques for assessing industrial hazards.

  (World Bank technical paper, ISSN 0253-7454 ; no. 55)
  Bibliography: p.
  1. Industrial safety--Handbooks, manuals, etc.
I. Technica, Ltd. (London, England)  II. Series.
T55.T36  1988     363.1'16     86-11135
ISBN 0-8213-0779-7

# Abstract

This manual, developed for use in assessing World Bank/IFC development proposals, provides guidelines for the identification of the potential hazards of new or existing plants or processes in the chemical and energy industries, and for the assessment of the consequences of the release of toxic, flammable or explosive materials to the atmosphere. It presents a structured, simplified approach for identifying the most serious potential hazards and for calculating their effect distances or damage ranges. It is the intention that by presenting a simplified approach, this manual should be amenable to use by engineers and scientists with little or no experience of hazard analysis. Further analysis with a view to mitigation of the hazards identified may be appropriate in many cases; at this stage it may be necessary to seek the advice of a specialist.

The basic procedure in a hazard analysis is: identify potential failures, calculate release quantities for each failure, and calculate the impact of each release on people and property. For large plants this can become highly complex, and therefore a simplified method is presented, in which the analysis has been divided into 14 steps. A spreadsheet technique has been devised to permit the analyses to be carried out on a programmable calculator or personal computer.

After the introductory material, this manual outlines the 14 steps which make up the hazard analysis. Chapters describing the details of the analysis follow. Chapter 3 covers choice of representative failures, and discusses the different forms which a release can take, all of which should be considered by the analyst. Chapter 4 describes the simplified models for calculating the consequences of a release; these include models for calculating discharge rates, jet dimensions, pool dimensions, dispersion behavior, and toxic, flammable or explosive impacts. Chapter 5 gives guidance on ordering and presenting the results for on-site and off-site hazards. Chapter 6 suggests ways in which the consequences, frequency and impact of the identified events may be reduced.

An appendix is included giving the "World Bank Guidelines on Identifying, Analysing and Controlling Major Hazard Installations in Developing Countries"; this suggests threshold quantities for plant inventories above which a major hazard assessment is recommmended, outlines what such an assessment should include, and gives guidance on emergency plans and restrictions on development in the vicinity of the installation.

# FOREWORD

The World Bank / International Finance Corporation (IFC) appraises and supervises the industrial developments which it sponsors. As part of these functions it is required to evaluate the hazard which a development might represent to the people and environment outside its boundaries. It is also required to evaluate the adequacy and effectiveness of the measures taken to control this hazard. To assist in this evaluation the Office of Environmental and Scientific Affairs has drawn up "The World Bank Guidelines for Identifying, Analysing and Controlling Major Hazard Installations in Developing Countries".

In order to implement these guidelines, it is necessary to carry out a hazard analysis of the proposed development to determine what damage could be caused by accidental releases of toxic, flammable or explosive materials from the development. This hazard analysis would identify the materials which are potentially hazardous and the incidents which could lead to their release. If any such incidents represent a major hazard to life or property, then efforts should be made to reduce the damage which the incident could cause. This could be done by introducing process changes or alternative processes, by reducing the inventories of hazardous materials, by providing robust secondary containment systems, by modifying site layouts, by moving to a different site or by improving control and management techniques.

If it is not possible to reduce the potential for damage by these methods, the next step might be to carry out a risk analysis. This risk analysis would calculate the probability of the hazardous incident occurring, and determine whether this probability could be reduced by changing such things as the manufacturing process, the safety systems, or the procedures for training, testing or maintenance. Ultimately, the hazard and risk analyses might show that the proposed combination of process and site represents an unacceptable threat to the neighbouring community and therefore a new site must be found.

This process of hazard and risk assessment can be applied to existing operations as well as to designs for change or expansion.

This manual aims to provide, in simplified form, the latest techniques used in the chemical industry to assess the consequences of releasing toxic, flammable or explosive materials into the atmosphere. A spreadsheet methodology has been devised to simplify hand calculations on scientific calculators, since the user of the manual might not have access to a computer. The spreadsheet methodology can also simplify computer applications of the techniques; however, more complex modelling procedures can be used when programming on micro-computers.

Although this manual has been prepared primarily for application to World Bank and IFC projects, the methodologies which it presents have wide application in the chemical industry, and others are welcomed and encouraged to use the manual. Further information concerning the environmental and the health and safety activities of the World Bank is available from:

Office of Environmental and Scientific Affairs
The World Bank
1818 H Street, N.W.
Washington D.C. 20433
U.S.A.

# CONTENTS

# LIST OF FIGURES

# LIST OF TABLES

# Chapter 1.

## Introduction

The chemical and energy industries use a wide variety of manufacturing, storage and control processes. These processes involve many different types of material, some of which can be potentially harmful if released into the environment, because of their toxic, flammable or explosive properties. A factor which contributes to the danger represented by these materials is that frequently they are not kept at atmospheric pressure and temperature: the processes used in the modern chemical and energy industries involve high pressures and temperatures; also, gases are frequently liqufied by refrigeration to facilitate storage in bulk quantities.

Under these circumstances, it is essential to achieve and maintain high standards of plant integrity through good design, management and operational control. Given the large quantities of potentially hazardous materials which are handled daily without incident, it is clear that the controls and safeguards which have been developed by industry are usually entirely effective. However, accidents do occur and these can cause serious injury to employees or the public, and damage to property. Therefore, when assessing design and development proposals for plants which handle such materials, it is essential to identify potential hazards. The necessary steps can then be taken to reduce the hazard (by design) or the risk (by high operating standards, safety devices etc.).

In order to conduct a hazard analysis it is important to follow a structured approach. In addition, the calculation methods used should be straight-forward and reliable. In the initial stages of a hazard analysis it is appropriate to apply simplified techniques in order to identify the most serious potential hazards; more sophisticated techniques can then be used to assess methods of reducing the hazards. This manual provides the minimum basis for any hazard assessment: it describes the framework necessary for a structured hazard analysis, and gives simplified formulae for calculating effect distances or damage ranges. The techniques given here have been applied to a variety of petrochemical and process plants and have been found to be effective.

These techniques should be used in conjunction with other methods of safety assessment, as appropriate for the plant being analysed. Some other methods are Hazard and Operability Studies (HAZOP), and Failure Modes and Criticality Effect Analysis (FMECA). The analyst might also use other methods of identifying and ranking potential hazards, such as the Dow Index or the ICI Mond Index. All these methods contribute to safety in the chemical industries but they are beyond the scope of this manual.

# Chapter 2.

## The Structure of a Hazard Analysis

## 2.1  BACKGROUND

The basic procedure in a hazard analysis of a chemical plant is as follows:  identify potential failures;  calculate the quantity of hazardous material released in each failure;  finally, calculate the impact of each release on plant equipment, people, the environment, and property.  This procedure can be applied to an entire plant or to part of a plant.

Applying the procedure to a large plant is very complex and difficult, and some simplifications must be made.  Some of these simplifications are widely-used techniques for structuring the hazard analysis efficiently:  this manual advises on how to sub-divide the plant into manageable units, how to avoid unnecessary calculations, and how to collate large quantities of results.  Other simplifications put the calculations within the range of an engineer who has only a programmable calculator or a micro-computer.  In addition,  many of the subtleties which would be found in a rigorous hazard analysis have been left out, simplifying the methods still further.

The hazard analysis of a complete plant has been divided into 14 main steps, as described in Section 2.2 and shown in Figure 2.1.  Before embarking on any analysis the user should read the entire manual and be confident that he or she understands the calculation methods and their limitations.

## 2.2  DESCRIPTION OF STEPS

### Step 1 - Divide the Site into Functional Units

Each unit should include at least one major storage vessel or pipe containing a hazardous material. The boundaries of each unit should be where there is means of isolating the vessel or pipeline from all other units in the event of a leak. Suitable means of isolation would be an emergency shut-down valve operated automatically, or a control valve which would be closed if the pressure or level in the vessel were falling. Manually operated valves would not be considered suitable unless they were remotely operated on a clear and unambiguous signal.

The releases from a unit are usually considered to come from a single point; if the parts of a unit are widely separated it might be advisable to split the unit into sub-units.

### Step 2 - Divide the Units into Components

Each unit must be split into "building-block" components. These are pieces of equipment such as those listed in Section 3.2 and shown in Figures 3.1 to 3.10. If the analyst is confronted with a component which is not included in the list, the listed item which most nearly corresponds with the component should be chosen so that the analysis can proceed.

### Step 3 - Find the Inventories of Hazardous Material in the Components

The inventories of all hazardous materials should be found by consulting Process Flow, and Piping and Instrumentation diagrams. The description of each inventory should include material type, phase, pressure, temperature, and volume or mass.

### Step 4 - Rank the Components by Inventory

The amount of calculation can be reduced to manageable proportions if the analysis is limited to the components which contain significant inventories. For a hazard assessment concerned with the potential on-site consequences of accidents it is difficult to quote minimum inventories. However, for off-site consequences reference can be made to the World Bank Guidelines in Appendix B; this lists the minimum inventories for different materials, i.e. the inventories above which a hazard assessment is considered necessary. It should be noted that potentially hazardous quantities may range from hundreds of grams to hundreds of tonnes depending upon the flammability or toxicity of the material. As a general rule, however, vapor releases can usually be ignored in the assessment of acute off-site risks if the vapor pressure within the vessel is less than 1 bar gauge.

### Step 5 - Find Representative Failure Cases for the Components

Only a small number of failure cases need be considered for each vessel, component, and pipe. A guide to the most commonly used failure cases is given in Figures 3.1 to 3.10, which show the components which are considered as "building blocks". These failure cases are based on conservative assumptions.

## Step 6 - Cluster the Release Cases

Some of the releases considered in a hazard assessment might involve the same material under similar conditions escaping through holes of similar sizes, though at different locations in the plant. To reduce the amount of calculation needed, these similar releases can be grouped together, or "clustered"; only one calculation is then required for each group.

## Step 7 - Calculate the Release Rates

The failures can be followed by an instantaneous or a continuous release of hazardous material. The quantity or rate of this release is calculated using the models described in Section 4.1. Figures 3.11 to 3.15 provide guidance on selecting the correct model to use; the choice of models depends on the nature of the material and the assumed discharge condition.

## Step 8 - Cluster the Release Rates

In order to reduce further the amount of calculation required, those releases which involve similar release rates (or similar amounts of a material ) at similar temperatures can also be clustered together. Dispersion and consequence calculations need be carried out only once for each group of clustered releases.

## Step 9 - Calculate the Consequences

On-site and off-site consequences are calculated using the models described in Sections 4.2 to 4.6; these give methods for estimating spreading or expansion, dispersion, fires, explosions, and toxic impacts. Guidance in selecting the correct models to use is given in Figures 3.11 to 3.15.

## Step 10 - Collate the Results

Chapter 5 includes a form which is designed to assist the analyst in recording, ordering and collating the results of a hazard assessment for a complex plant.

## Step 11 - Plot Effect Distances

Ultimately, the results of the hazard assessment calculations should be considered in relation to the local geography and population. Since the results for each release case include an "effect distance", hazard impacts can be estimated by drawing "effect radii" as circles on local maps.

## Step 12 - Estimate Event Frequencies

The analyst can use reliabilty data to estimate the frequency of occurrence for each failure case. If failure data for the plant under examination exist, these should be used in preference to more general failure data. At this stage the analyst will be able to make only a superficial estimate of frequencies; a full risk analysis would involve reliability and availability analyses, which are beyond the scope of this manual. However, the frequencies are important because they add an additional perspective to the analysis, and are useful when deciding how to allocate limited resources for remedial measures.

## Step 13 - Interpret the Results

The analyst should then decide whether the plant represents an unacceptable threat to its workers or to the community.

## Step 14 - Choose and Analyse Remedial Measures

If the hazard is unacceptable the analyst should consider ways of reducing it. Many methods have been developed to reduce hazards in complex process plants. The details of many of these methods will be specific to the plant in question so this manual cannot provide a comprehensive description of all the options which might be available. However, some suggestions and examples are given in Chapter 6. The analyst can quantify the benefits of a remedial measure by repeating the relevant consequence calculations.

**FIGURE 2.1 :** *Structure of a Hazard Analysis*

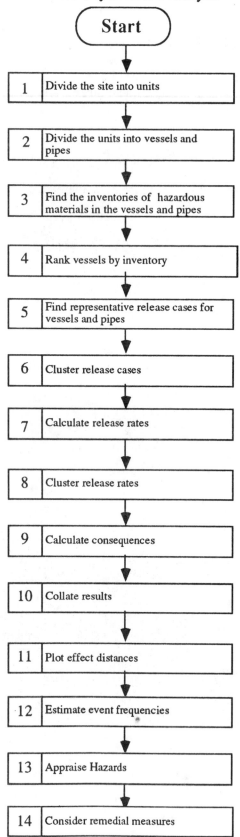

Start

| 1 | Divide the site into units |
| 2 | Divide the units into vessels and pipes |
| 3 | Find the inventories of hazardous materials in the vessels and pipes |
| 4 | Rank vessels by inventory |
| 5 | Find representative release cases for vessels and pipes |
| 6 | Cluster release cases |
| 7 | Calculate release rates |
| 8 | Cluster release rates |
| 9 | Calculate consequences |
| 10 | Collate results |
| 11 | Plot effect distances |
| 12 | Estimate event frequencies |
| 13 | Appraise Hazards |
| 14 | Consider remedial measures |

# Chapter 3.

## Failure Cases

## 3.1 BACKGROUND

This chapter describes the steps involved in a consequence analysis. It describes the decisions which the analyst must make at different stages, but does not describe the consequence models and calculations; these are given in Chapter 4.

The first step in an analysis is to select representative release cases. Release cases are accidents or failures which lead to a release of hazardous material. Since the number of such failures is very large, it is usual to consider only a few release cases, chosen as representative of the complete range of release cases. The manual lists 10 items which are the "building blocks" of process plants, and suggests the release cases which should be considered for each item.

The next step in the analysis is to determine the properties and state of the materials released, considering both normal and abnormal plant operation. With this information the analyst can select the appropriate set of consequence calculations.

The final step is to carry out the consequence calculations. There are many possible sequences of events following an accident or failure and each sequence requires a different set of consequence models. Diagrams, called "event trees", which describe typical sequences are included in the chapter in order to help the analyst choose the appropriate sequence.

## 3.2 RELEASE CASES

The first step in establishing a representative set of release cases is to list the components of the plant. Only a few different types of process plant component are of importance in hazard analysis and a simplified analysis can usually be limited to 10 types. These 10 components are shown in Figures 3.1 to 3.10. They are:

| | |
|---|---|
| Pipes | Figure 3.1 |
| Flexible connections | Figure 3.2 |
| Filters | Figure 3.3 |
| Valves | Figure 3.4 |
| Pressure / process vessels | Figure 3.5 |
| Pumps | Figure 3.6 |
| Compressors | Figure 3.7 |
| Storage tanks    (i.e.  ambient conditions) | Figure 3.8 |
| Storage vessels  (i.e.  pressurised or refrigerated) | Figure 3.9 |
| Flare / vent stack | Figure 3.10 |

Although a plant might contain variants of a particular component, the variations are usually small enough to make it easy to decide to which group a piece of equipment belongs.

The next step in establishing a set of release cases is to choose representative failures for each component. Figures 3.1 to 3.10 suggest typical failure modes; for example, for a pipe, typical failure modes are pipe leaks, flange leaks, and weld failures. The figures also give representative failure sizes for each failure mode; for example, for a pipe leak the suggested sizes are 20% and 100% of the pipe diameter. The analyst should consider whether the failure modes and sizes are appropriate to the items on the plant.

The sets of failures given here are the minimum capable of representing the items in a hazard analysis and the analyst might decide to consider additional modes and sizes. For example, human error could be included as a failure mode for some types of equipment: a valve could be accidentally left open, or a storage tank could be over-filled.

As well as considering releases arising from a breach of containment, the analyst should consider abnormal releases from equipment which is designed to release material, e.g. flares, vent stacks, and drains. Such releases are mentioned briefly in Figure 3.10 and are discussed more fully below.

Many plants rely on vent, scrubbing and flare systems for the safe handling of discharges and upset conditions. In some circumstances these systems might not be adequate and hazardous materials might be discharged; the analyst should include such discharges in the list of release cases. Some of the circumstances which can lead to a hazardous release are given below:

- Flow rate into the discharge system is greater than the flow for which the system was designed. This might mean that the material is not scrubbed properly, or that the equipment ruptures, releasing the material in a dangerous location.

- A flammable material is discharged into a cold vent system which is not designed to handle flammables.

- A flammble material fails to ignite at the flare stack.

- An unstable or abnormal reaction produces hazardous materials for which the discharge system was not designed.

FIGURE 3.1 :                    PIPE

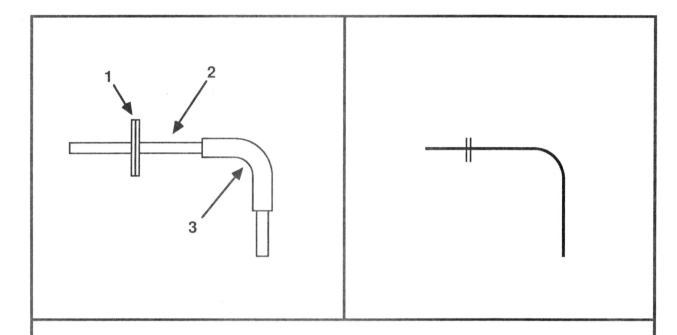

Includes:

Pipes, Flanges, Welds, Elbows.

| Typical failures | Suggested Failure Sizes |
|---|---|
| 1. Flange leak | 20% pipe diameter |
| 2. Pipe leak | 100% and 20% pipe diameter |
| 3. Weld failure | 100% and 20% pipe diameter |

FIGURE 3.2:  **FLEXIBLE  CONNECTIONS**

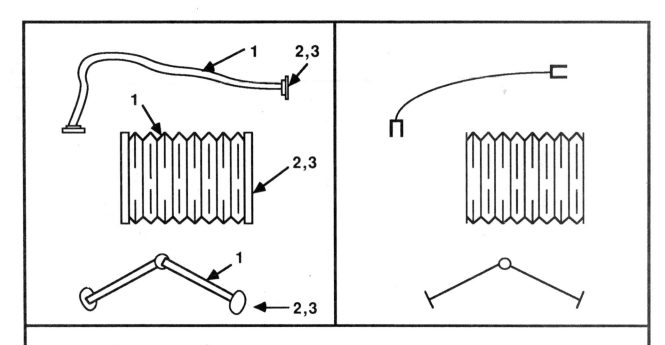

Includes:

Hoses, Bellows, Articulated Arms.

| Typical failures | Suggested Failure Sizes |
|---|---|
| **1.** Rupture leak | 100% and<br>20% pipe diameter |
| **2.** Connection leak | 20% pipe diameter |
| **3.** Connection mechanism<br>failure | 100% pipe diameter |

**FILTER**

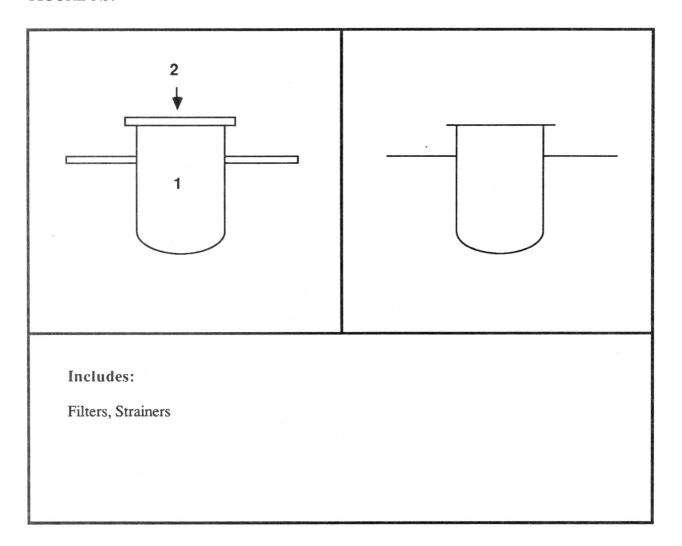

Includes:

Filters, Strainers

| Typical failures | Suggested Failure Sizes |
|---|---|
| 1. Body leak | 100% and 20% pipe diameter |
| 2. Pipe leak | 20% pipe diameter |

VALVE

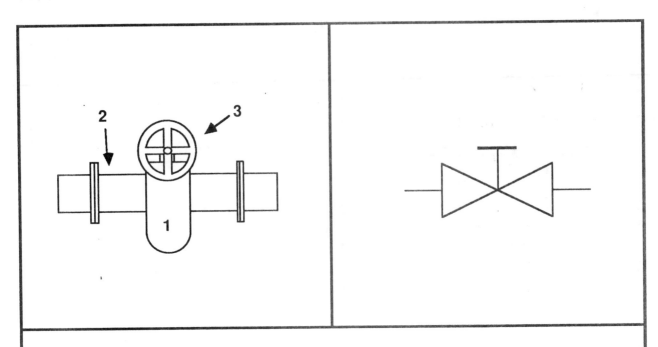

**Includes:**

Ball, Gate, Globe, Plug, Needle, Butterfly, Choke, Relief, ESD-valves

| Typical failures | Suggested Failure Sizes |
| --- | --- |
| 1. Housing leak | 100% and 20% pipe diameter |
| 2. Cover leak | 20% pipe diameter |
| 3. Stem failure | 20% pipe diameter |

FIGURE 3.5:         PRESSURE VESSEL / PROCESS VESSEL

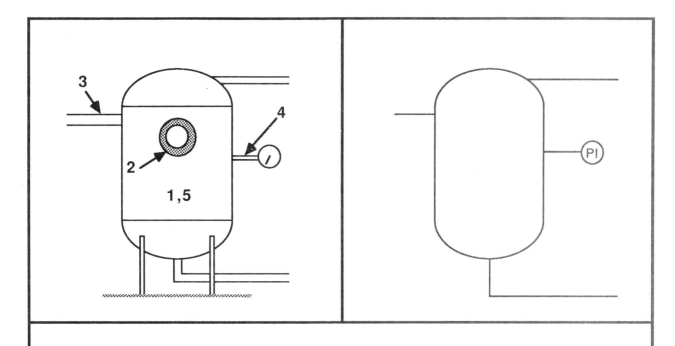

**Includes:**

Separators, Scrubbers, Contactors, Reactors, Heat Exchangers,  Fired Heaters, Columns, Pig Launchers / Receivers, Reboilers.

| Typical failures | Suggested Failure Sizes |
|---|---|
| **1.** Vessel rupture;<br>Vessel leak | Total rupture;<br>100% pipe diameter of largest pipe |
| **2.** Manhole cover leak | 20% opening diameter |
| **3.** Nozzle failure | 100% pipe diameter |
| **4.** Instrument line failure | 100% and<br>20% pipe diameter |
| **5.** Internal explosion | Total rupture |

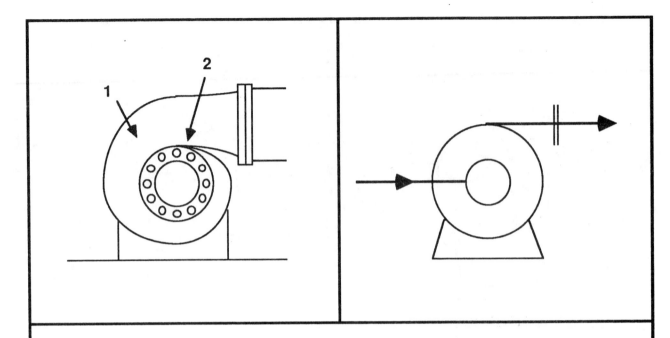

**Includes:**

Centrifugal pumps, reciprocating pumps

| Typical failures | Suggested Failure Sizes |
|---|---|
| 1. Casing failure | 100% and 20% pipe diameter |
| 2. Gland leak | 20% pipe diameter |

COMPRESSOR

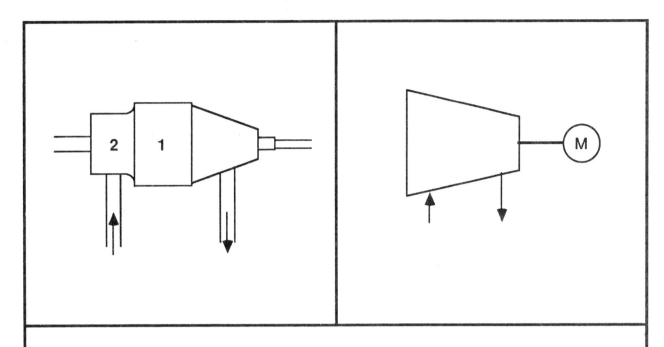

Includes:

Centrifugal Compressors, Axial Compressors, Reciprocating Compressors

| Typical failures | Suggested Failure Sizes |
|---|---|
| 1. Casing failure | 100% and 20% pipe diameter |
| 2. Gland leak | 20% pipe diameter |

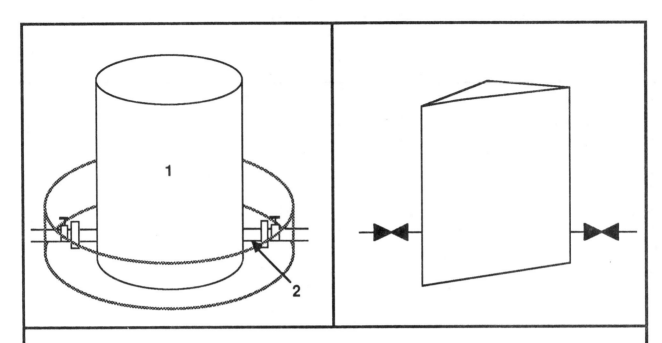

**Includes:**

All tanks at ambient conditions
(The pipe connections and bund wall are also considered to be part of this component )

| Typical failures | Suggested Failure Sizes |
|---|---|
| 1. Vessel failure | Total Rupture |
| 2. Connection leak | 100% and 20% pipe diameter |

## FIGURE 3.9: STORAGE VESSEL (PRESSURISED or REFRIGERATED)

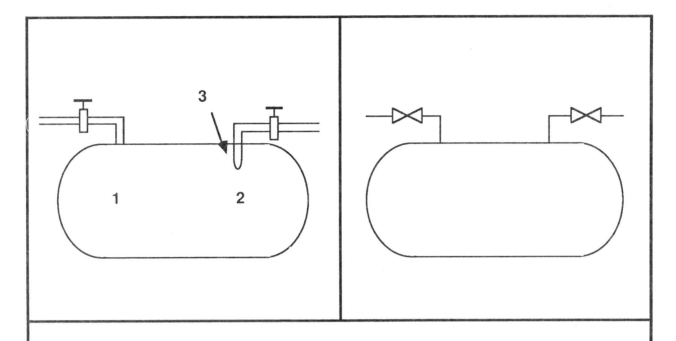

Includes:

Pressurised Storage or Transport Vessels, Refrigerated Storage or Transport Vessels, Buried or Non-buried Vessels.

| Typical failures | Suggested Failure Sizes |
|---|---|
| 1. BLEVE (non-buried case only) | Total rupture (ignited) |
| 2. Rupture | Total rupture |
| 3. Weld failure | 100% and 20% pipe diameter |

Note :  These storage vessels may have bund walls which should be taken into consideration in the analysis.

**FLARE / VENT STACK**

**Includes:**

All Flares or Vent Stacks.
(The manifold, vent scrubber and knock-out drum are also considered
   to be part of this component )

| Typical failures | Suggested Failure Sizes |
|---|---|
| **1.** Manifold / Drum leak | 100% and 20% pipe diameter |
| **2.** Discharge beyond specification | Should be estimated. |

## 3.3 PROPERTIES OF RELEASED MATERIALS

Once the release cases have been defined, the analyst should determine the properties of the material released in each case. The properties relevant to a hazard analysis are:

Phase (i.e. liquid, gas or two-phase)
Pressure
Temperature
Flammability
Toxicity

Out of the wide range of different combinations of these properties a few classes of combinations, or "categories", are especially important in hazard analysis. The following categories are used in this manual:

a)  Liquid at ambient pressure and temperature
    ("liquid, ambient pressure and temperature")

b)  Liquefied gas under pressure
    ("liquid, pressurised")

c)  Liquefied gas at low temperature
    ("liquid, refrigerated")

d)  Gas under pressure
    ("gas, pressurised")

The above four categories cover most releases. However, there are some exceptions, two of which are discussed below.

*Boiling Liquid Expanding Vapour Explosion (BLEVE)*

A BLEVE can occur if a pressurised tank of flammable liquid is subjected to a fire. The heat input to the tank has two effects: first, it raises the internal pressure; second, it weakens the tank shell, especially in the top parts, which are not wetted internally. These effects can cause the tank to burst catastrophically, releasing a large quantity of liquid which vaporises violently and ignites to form a fireball. A BLEVE will also produce large projectiles from parts of the ruptured vessel or from neighbouring equipment; the projectiles can cause significant damage.

*Mixtures of Hazardous Materials*

Such releases are not easy to model, but approximate methods have been developed for some mixtures:

-  If the mixture consists mainly of one component, the physical properties of that component can be used for the whole mixture.

-  If only one component of the mixture is toxic, the toxicity values for this diluted component can be applied in conjunction with the physical properties of the mixture; so a leak of butane with a trace of hydrogen sulphide can be modelled using the physical properties of butane to calculate release rates and dispersion, and the toxicity values of dilute hydrogen sulphide to calculate the toxic impact.

## 3.4   CONSEQUENCES OF A RELEASE

A release can have many different consequences and the analyst should try to consider all of them. This is a complex task and needs a structured approach.   This manual provides aids to such an approach in the form of a "Failure Case Definition Tree" (Figure 3.11) and four "event trees" (Figures 3.12 to 3.15).   These figures are called trees because they have a branching structure which describes alternative sequences of events.   At each node, or branching point, the analyst is asked a question about the release;  the answer determines which path the analyst should then follow.   For all of these trees, if the analyst considers that both answers are possible,  then  both paths should be followed.

The Failure Case Definition Tree is simple;  it helps the analyst choose the appropriate event tree by asking questions about the properties of the release.

The four event trees are more complicated than the Failure Case Definition Tree.   They are:

| | |
|---|---|
| Flammable Gas Event Tree | Figure 3.12 |
| Toxic Gas Event Tree | Figure 3.13 |
| Flammable Liquid Event Tree | Figure 3.14 |
| Toxic Liquid Event Tree | Figure 3.15 |

They trace the possible sequences of events after a release by asking such questions as:  "Is there immediate ignition?";  "Does a dense cloud form?".   At each branch of the tree the analyst is directed to the section in Chapter 4 which explains the appropriate consequence calculations.   Each event tree is discussed further in Sections 3.4.1 to 3.4.3.   A question which is very important in all release cases is "Are shut-down and isolation  successful ?"  This is not included in the event trees but the issues of isolation and shut-down are dealt with separately in Section 3.4.4 since they are common to all four trees.

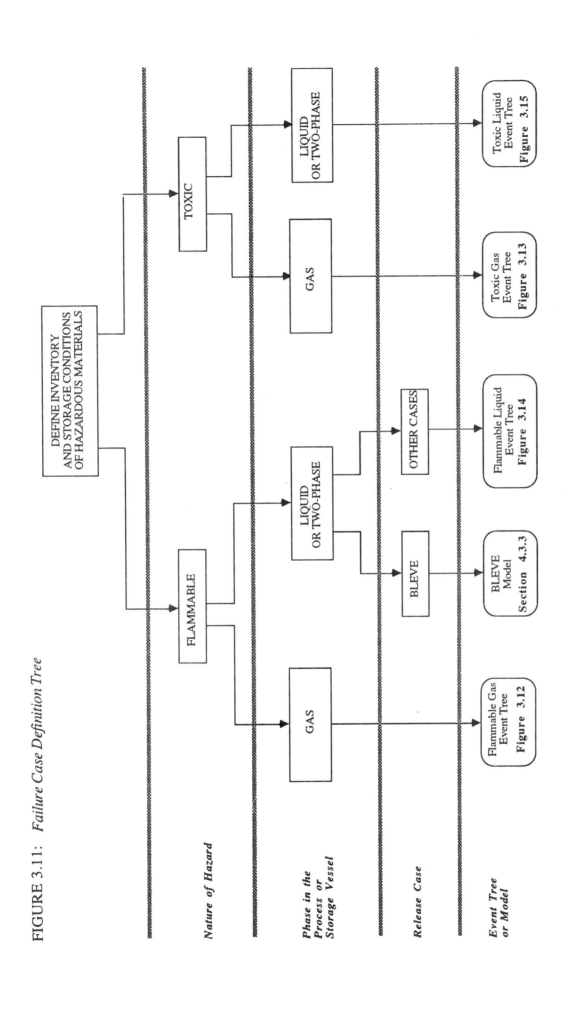

FIGURE 3.11: *Failure Case Definition Tree*

## 3.4.1  Flammable Gas Event Tree  (Figure 3.12)

Since flammable gases are usually only a danger if they ignite, identifying the sources and probabilities of ignition is a very important part of the analysis.  It is convenient to divide ignition into two categories according to the timing of the ignition, as follows:

### a) immediate ignition
In this case the gas is ignited while it is still escaping from containment.  Immediate ignition  prevents a large cloud of vapor developing, but can result in a jet flame or a fireball, depending on the nature of the release.  These can cause damage in the immediate vicinity of the release but rarely affect anything outside the plant boundary.

### b) delayed ignition
This occurs after the material has escaped from containment and has formed a cloud which is drifting downwind.  Delayed ignition can result in an explosion or a flash fire, which can cause widespread damage.

The heat and pressure effects calculated for fires or explosions are used to assess fatalities and material damage, and also to determine whether there are likely to be any "knock-on" (or "domino") effects, such as damage to another piece of equipment which contains hazardous materials.  The releases arising from knock-on effects should be analysed like any other releases.

Another important part of the analysis of a gas release is the calculation of the density of the cloud, since density is a major factor in determining how far the cloud travels and spreads before it is diluted down to a safe concentration.  The choice of the appropriate model for dispersion is made according to the density of the cloud relative to the ambient air.

## 3.4.2  Toxic Gas Event Tree  (Figure 3.13)

A toxic gas release is simpler to analyse than a flammable gas release since ignition need not be considered.  Selecting the cloud dispersion model is the most important part of the analysis; as with a flammable gas, the density of the cloud determines the appropriate model.  In this manual greatest emphasis is given to the dispersion of clouds which are denser than air since the area under these clouds can be large, and such clouds can cover great distances before they are diluted to a safe concentration.  For these reasons, the effects of dense clouds can be widespread.

### 3.4.3 Liquids Event Trees (Figure 3.14 and 3.15)

Liquids are fairly easy to contain and liquid fires usually affect only a small area. Gases, on the other hand, are very difficult to contain and can affect a very wide area. Therefore, a liquid release which remains a liquid is unlikely to represent a serious hazard. However, many liquids will vaporise after release and the gas which is produced can represent a serious hazard and should be analysed using the appropriate gas event tree. Therefore, the most important consideration in a liquid release is the amount of gas which evaporates. This depends on the ambient temperature and on the storage conditions of the liquid. The three liquid categories listed in Section 3.3 will each produce gas at a different rate, as described below:

a) **liquid, ambient pressure and temperature**
   Such a liquid will form a pool and will slowly vaporise because of the advection of wind over its surface .

b) **liquid, pressurised**
   Some of the liquid will flash to vapor immediately upon release, and the remaining liquid will form a pool which will vaporise as it absorbs heat from its surroundings. The proportion of the liquid vaporising depends on the properties of the material and on ambient temperature; some releases will vaporise completely upon release and will not form a pool.

c) **liquid, refrigerated**
   Such a liquid will form a pool, which will absorb heat from its surroundings, and evaporate to give a gas cloud. The rate of evaporation is likely to be lower than for a pressurised release, but higher than for a release of liquid from ambient conditions.

The above three categories are found in both flammable and toxic releases and for both types of release the most important aspect of the analysis is the calculation of a vapor release rate for use with the appropriate Gas Event Tree. However, the Flammable Liquids Events Tree is more complex than the Toxic Liquids Event Tree because it also considers ignition of the escaping material and of the pool.

### 3.4.4 Isolation of a Release

The behaviour of a release can be very dependent on its duration. The duration depends on the amount of material available to be released, which in turn depends on the speed and effectiveness of shut-down or isolation. Therefore, isolation can affect the consequences of a release and it is important to make a realistic estimate of the time required for isolation. This time will depend on the following:

a) **Leak Detection**
   It is usual to assume that major ruptures and leaks will be detected immediately, either by instruments or by operators. Smaller leaks should be detected by gas or flame detectors; the analyst should determine the position and reliability of such detectors.

b) **Shut-down Activation**
   The speed of shut-down will depend on whether actuation is manual or automatic. The response time for manual actuation depends on alarm design, operating procedures and operator training; response times of 3 to 15 minutes are typical.

## c) Shut-down Valves

The closing time of a shut-down valve will depend on the valve size and the pressures involved; a typical closing time for a large, high-pressure valve is 30 seconds. It is possible that the valve will not close at all, so the analyst should determine the availability and reliability of the shut-down valves on the plant.

# FIGURE 3.12: *Flammable Gas Event Tree*

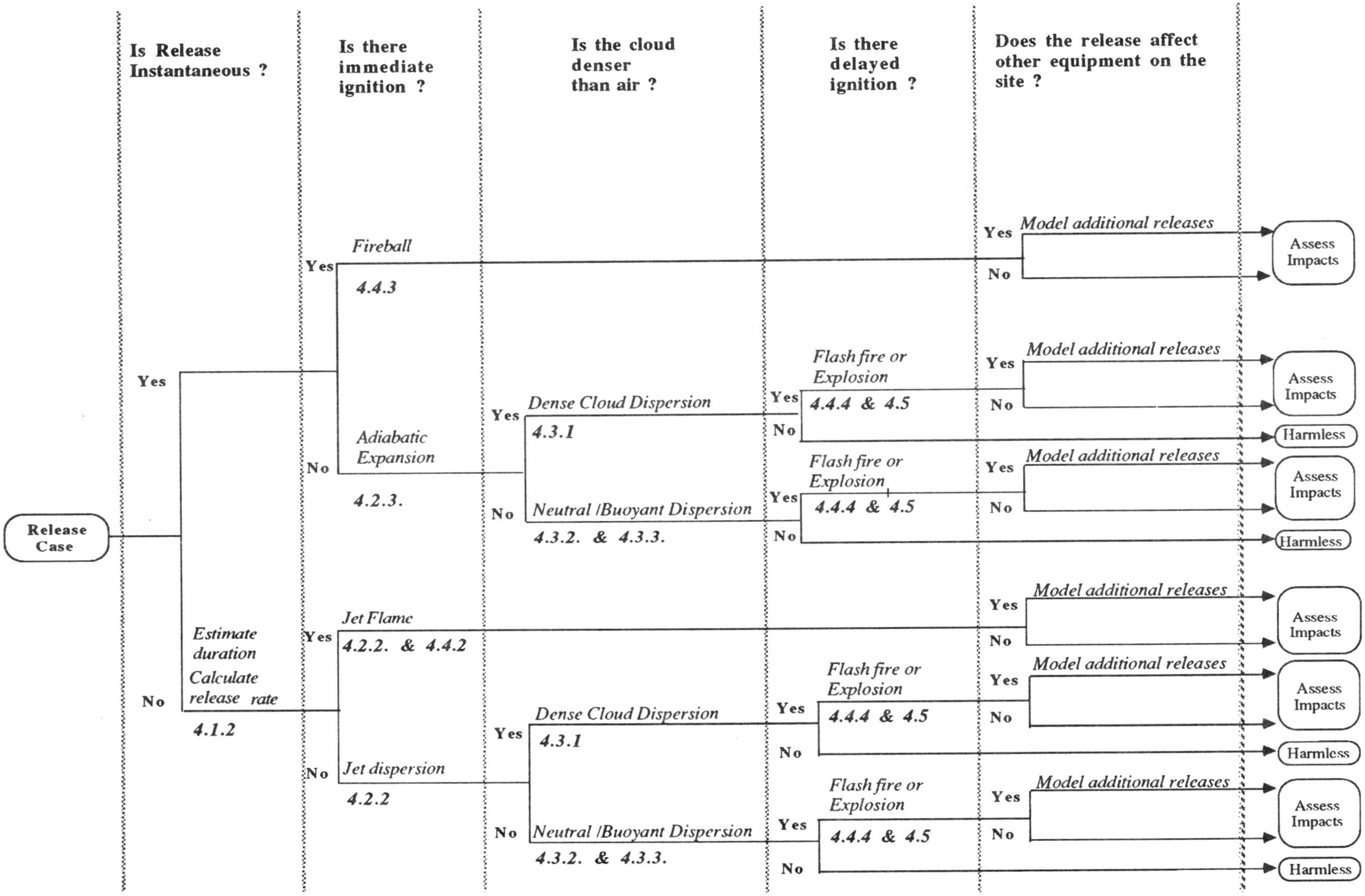

**Is Release Instantaneous ?** — **Is there immediate ignition ?** — **Is the cloud denser than air ?** — **Is there delayed ignition ?** — **Does the release affect other equipment on the site ?**

**Release Case**

**Yes** branch:
- **Yes** — *Fireball* **4.4.3** → **Yes** *Model additional releases* → Assess Impacts / **No** → Assess Impacts
- **No** — *Adiabatic Expansion* **4.2.3.**
  - **Yes** — *Dense Cloud Dispersion* **4.3.1** → **Yes** *Flash fire or Explosion* **4.4.4 & 4.5** → **Yes** *Model additional releases* → Assess Impacts / **No** → Assess Impacts / **No** → Harmless
  - **No** — *Neutral /Buoyant Dispersion* **4.3.2. & 4.3.3.** → **Yes** *Flash fire or Explosion* **4.4.4 & 4.5** → **Yes** *Model additional releases* → Assess Impacts / **No** → Assess Impacts / **No** → Harmless

**No** branch — *Estimate duration Calculate release rate* **4.1.2**:
- **Yes** — *Jet Flame* **4.2.2. & 4.4.2** → **Yes** *Model additional releases* → Assess Impacts / **No** → Assess Impacts
- **No** — *Jet dispersion* **4.2.2**
  - **Yes** — *Dense Cloud Dispersion* **4.3.1** → **Yes** *Flash fire or Explosion* **4.4.4 & 4.5** → **Yes** *Model additional releases* → Assess Impacts / **No** → Assess Impacts / **No** → Harmless
  - **No** — *Neutral /Buoyant Dispersion* **4.3.2. & 4.3.3.** → **Yes** *Flash fire or Explosion* **4.4.4 & 4.5** → **Yes** *Model additional releases* → Assess Impacts / **No** → Assess Impacts / **No** → Harmless

FIGURE 3.13: *Toxic Gas Event Tree*

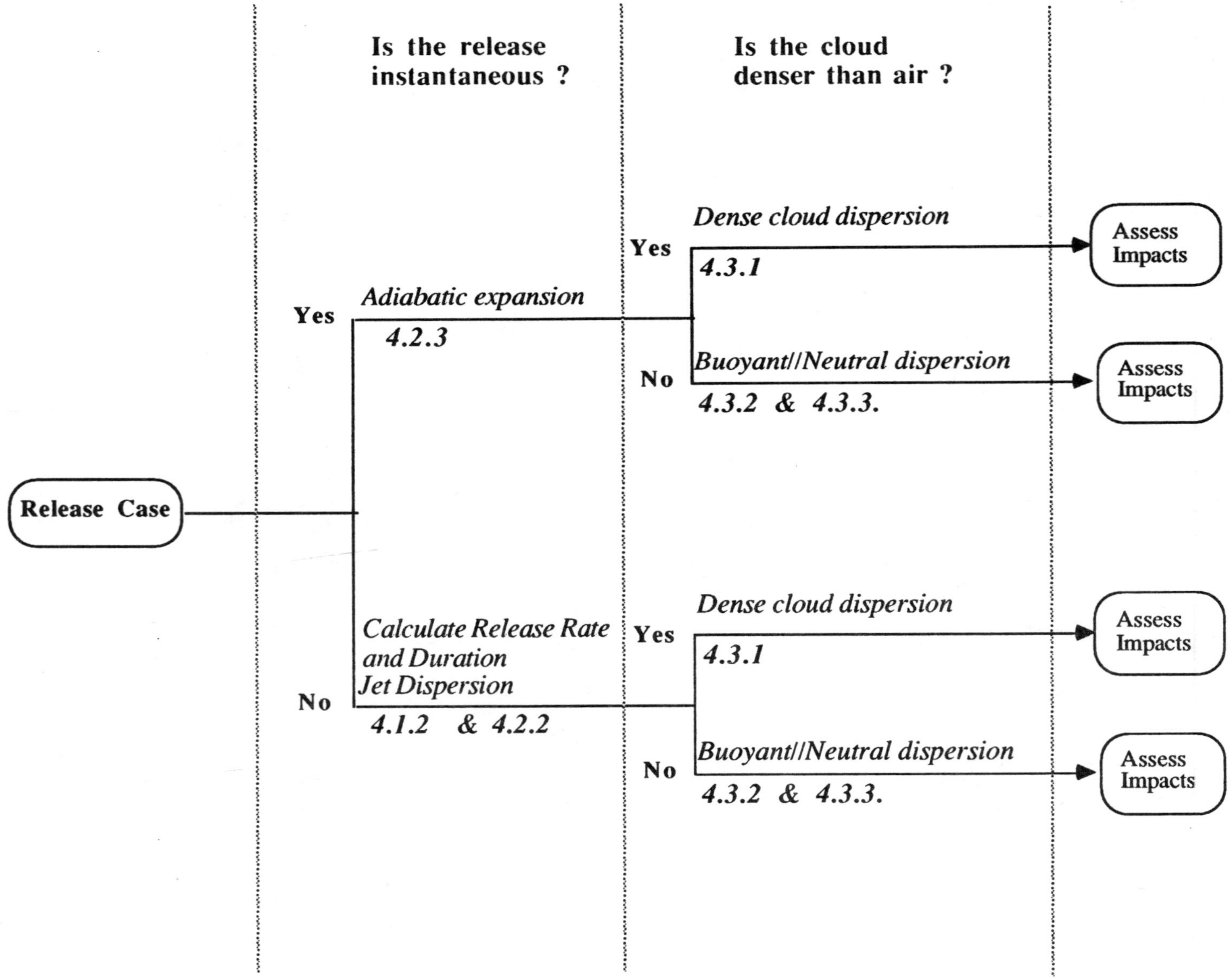

FIGURE 3.14: *Flammable Liquid Event Tree*

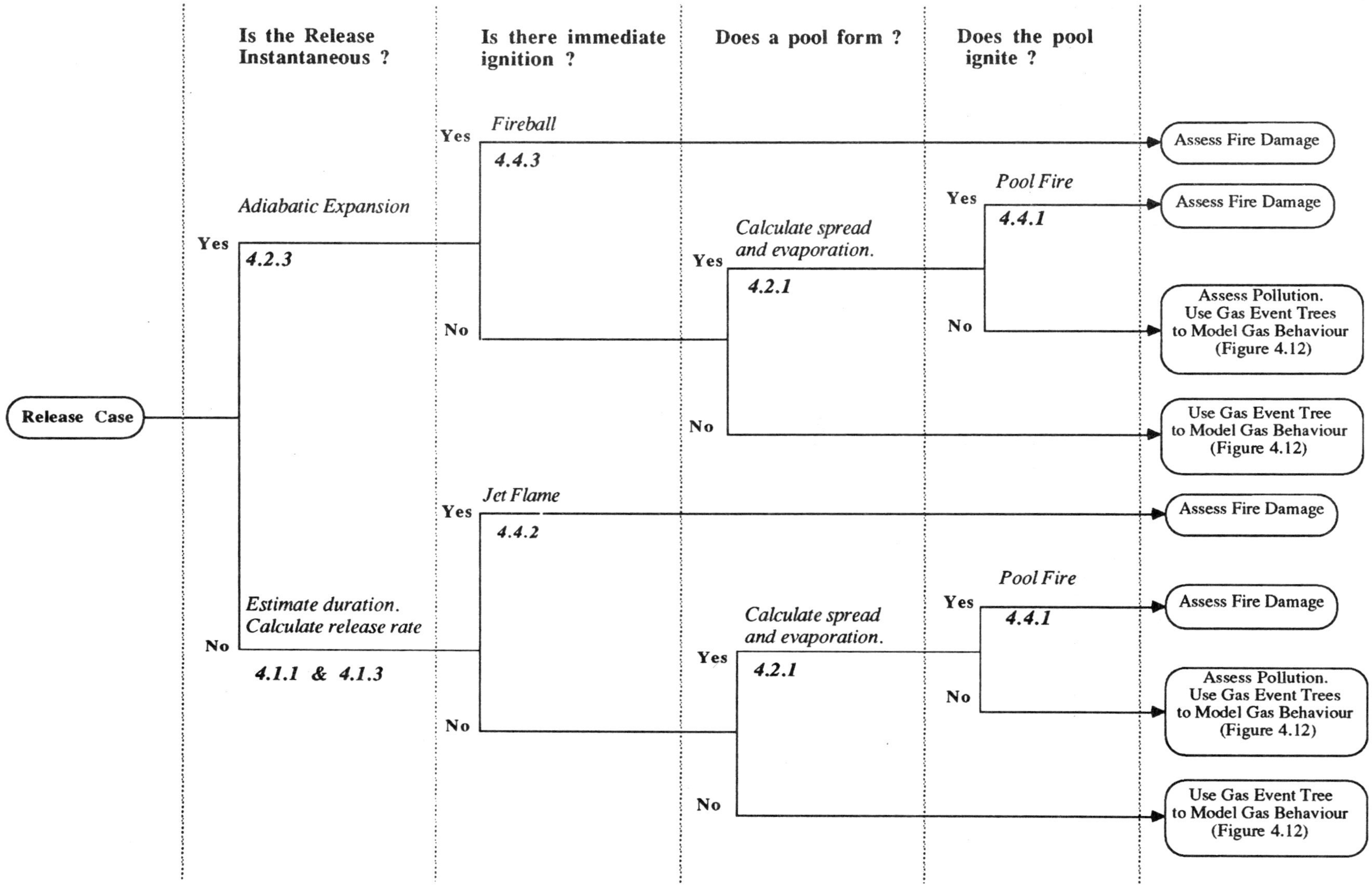

FIGURE 3.15: *Toxic Liquid Event Tree*

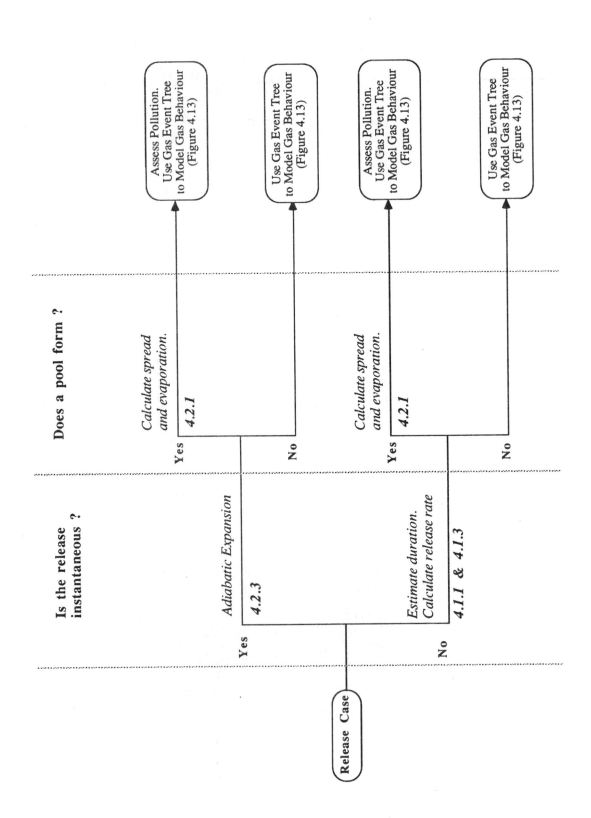

# Chapter 4.

## Consequence Calculations

This chapter gives the theoretical basis for estimating discharge rates, dispersion behaviour, and toxic or flammable impacts of the postulated failure cases. Wherever possible, the equations have been simplified to enable an analyst with only a programmable calculator to conduct an initial hazard assessment. The nomenclature and notation used in this chapter are summarised in Appendix A.

Throughout this chapter the units are SI units unless otherwise stated.

## 4.1 OUTFLOW CALCULATIONS

Most incidents start with a hazardous material escaping from confinement, owing to either failure of containment, or an abnormal discharge from an engineered outlet such as a relief valve. There are well-known equations which can be used to calculate the rate of release given the size of the hole, and the thermodynamic and physical properties of the escaping material. Judgement may be required where irregular holes or non-uniform conditions are involved. The problem of the irregular hole is usually solved by using an equivalent hole diameter for the net open area. The most common type of non-uniform condition is a changing pressure and the analyst should consider what is happening to the pressure in the container over the duration of the release; before the vessel is isolated it is still supplied with material and thus maintains a relatively high pressure; after isolation the pressure in the container and the release rate will fall, and the analyst should try to take this into account.

This section deals with three types of discharge:

**Liquid Release** : the release of a liquid from containment; the fluid remains liquid while it is flowing through the hole although it might flash subsequently

**Gas Release** : the discharge of a gas from containment

**Two-Phase Release** : the release of a mixture of gas and liquid (usually resulting from boiling of the liquid under discharge conditions).

## 4.1.1      Liquid Outflow

**METHOD**     Use of Bernoulli flow equation to calculate release rate of hazardous liquid from containment. Use of adiabatic flash calculation to determine liquid/vapor ratio immediately after discharge.

**OUTPUT**     Discharge rate and fraction of liquid which flashes off immediately after discharge.

**CONSTRAINTS**     The liquid should not flash inside the orifice.

---

**The Method**     Liquid release rates are calculated using the Bernouilli flow equation. The release rate for a liquid outflow from containment is given by:

$$Q = C_d A_r \rho_1 \sqrt{ \frac{2 (P_1 - P_a)}{\rho_1} + 2gh }$$

For liquid which is stored at ambient pressure the driving force is the liquid head. For liquid which is stored under pressure the driving forces are the liquid head and the difference between the vapor pressure of the liquid, $P_1$ and atmospheric pressure, $P_a$.

The above equation assumes that the liquid remains liquid while it is escaping, i.e. it does not flash to form a vapor while it is passing through the hole. This is assumed to be the case for releases from the liquid space of a vessel or pipe where the length / diameter ratio of the hole is small (i.e. <12).

However, if the liquid is super-heated it will flash once it has escaped and is at atmospheric pressure. The heat required for vaporisation is taken from the liquid itself so that any liquid which is left will have been cooled to its atmospheric boiling point. The fraction of liquid which will flash immediately is given by :

$$F_{vap} = \frac{C_{P_1} (T_1 - T_b)}{H_{vap_b}}$$

Almost invariably, $F_{vap}$ as calculated by this equation lies between 0 and 1. In this case some liquid will remain in the cloud as a finely dispersed aerosol. Some of this liquid will evaporate as air at ambient temperature mixes with the liquid spray. If the heat transfer from the air is not sufficient to vaporise all of the liquid, some of the liquid will drop to the ground and form a pool.

There is no accepted model for whether or not rain-out of liquid occurs. However, spill experiments have shown that the formation of a pool is unlikely if $F_{vap}$ is greater than about 0.2. It could be assumed that for values of $F_{vap}$ less than 0.2, there is a linear relationship between $F_{vap}$ and proportion of liquid entrained: when $F_{vap} = 0$, none of the liquid is entrained; when $F_{vap} = 0.1$, 50% of the liquid is entrained, etc.

If $F_{vap}$ is greater than one, the liquid will evaporate completely before reaching atmospheric pressure to give a jet or cloud of vapor with a temperature which may be above $T_b$.

---

Outputs

This method gives the discharge rate of a liquid from refrigerated or pressurised containment. It is also used to calculate the fraction of liquid which flashes to atmosphere immediately after discharge.

Input

(i) Discharge coefficient, $C_d$.
This depends on the shape of the hole and on the phase of the flow. For liquid flow a value of 0.6-0.64 should be used.
(ii) Thermophysical data for the material, e.g. from DIPPR (1985).
(iii) Initial pressure and hydrostatic head.
(iv) Effective open area, $A_r$, for the hole.

Assumptions

This method is based on the Bernoulli equation which is a standard equation for liquid outflow. The main assumptions are the value of $C_d$ and the assumption of incompressible flow. This method yields instantaneous discharge rates and no allowance is made for the time dependency of the discharge as the pressure or liquid head falls.

Accuracy

This method for calculating discharge rates is considered to be as accurate as necessary for the type of study involved (i.e. better than + 5%). It should be noted that when applied to a discharge over a finite period of time, the results obtained are conservative if the initial pressure and liquid head are used throughout.

Application

This method can be applied to releases of liquids from enclosed systems with a hydrostatic liquid head, h, and an internal pressure, $P_1$.

Resources Required

This calculation can be performed with a calculator.

Linkage with Other Models

The release rate of liquid is used to model pool behaviour; the release rate of gas and entrained aerosol is used to model jet or cloud behaviour.

### 4.1.2      Gas Outflow

**METHOD**    Use of the gas flow equation for calculating discharge rates for gases from sources under pressure.

**OUTPUT**    Gas discharge rate.

**CONSTRAINTS**    The simple equations assume ideal gas behaviour which is probably reasonable for all but very high (near critical) pressures.

---

**The Method**    Discharges from vessels and pipes containing gas under pressure are normally readily calculated using standard equations for gas flow.

The first step in the calculations is to determine whether the flow is critical, i.e. sonic or choked, or sub-critical. The distinction is made as follows, assuming reversible adiabatic expansion and ideal gas behaviour:

Flow is **critical** if

$$P_a < P_1 \left[ \frac{2}{\gamma+1} \right]^{\frac{\gamma}{(\gamma-1)}}$$

Flow is **sub-critical** if

$$P_a > P_1 \left[ \frac{2}{\gamma+1} \right]^{\frac{\gamma}{(\gamma-1)}}$$

The second step is to calculate the discharge rate using the following formula taken from Crane (1981):

$$Q = Y C_d A_r P_1 \sqrt{ \left[ \frac{M\gamma}{R T_1} \right] \cdot \left[ \frac{2}{\gamma+1} \right]^{\frac{(\gamma+1)}{(\gamma-1)}} }$$

For **critical** outflow:

$$Y = 1.0$$

For **sub-critical** outflow:

$$Y = \left[\frac{p_a}{p_1}\right]^{\frac{1}{\gamma}} \cdot \left[ 1 - \left\{ \frac{p_a}{p_1} \right\}^{\frac{(\gamma-1)}{\gamma}} \right]^{\frac{1}{2}} \cdot \left[ \left\{ \frac{2}{\gamma-1} \right\} \cdot \left\{ \frac{\gamma+1}{2} \right\}^{\frac{(\gamma+1)}{(\gamma-1)}} \right]^{\frac{1}{2}}$$

The analyst should also consider whether the flow rate and duration of the release are sufficient to reduce the inventory of the vessel significantly. As the inventory falls so does the pressure and therefore also the flow rate. If the flow rate is small or the duration is short, it is usually acceptable to use the initial discharge rate in the consequence calculations; otherwise the analyst should choose an equivalent discharge rate for the whole duration.

---

| | |
|---|---|
| Outputs | This method gives the initial discharge rate for a pressurised gas release. |
| Inputs | (i) Discharge coefficient, $C_d$. For a gas release a value of 1.0 should be used. <br> (ii) Thermophysical data for the gas, e.g. from DIPPR (1985) and Crane (1981). <br> (iii) Initial pressure of release $P_1$. <br> (iv) An effective open area $A_r$, for the hole. |
| Assumptions | This method assumes reversible adiabatic expansion and ideal gas behaviour. |
| Accuracy | This method is sufficiently accurate for hazard assessment provided ideal gas behaviour is a reasonable assumption. |
| Application | This method can be applied to the discharge of gases from large storage vessels or pipes. |
| Resource Requirements | This calculation can be performed with a calculator. |
| Linkage with Other Models | The release rate of gas is used to model jet or cloud behaviour. |

## 4.1.3　　Two-phase Outflow

**METHOD**　　Fauske/Cude method to calculate rate of discharge from two-phase critical flows.

**OUTPUT**　　Discharge rate and fraction of liquid which flashes during discharge.

**INPUTS**　　Discharge coefficient and thermophysical data.

**CONSTRAINTS**　　Alternative methods are required for more complex situations; this section suggests some methods which yield approximate solutions.

---

**The Method**

Two-phase critical flows can occur in failures of connections to the vapor space of vessels which contain superheated liquids under pressure. They also occur in pipes containing superheated liquid if there is a great length of pipe upstream of the failure, since in this situation a fully-developed critical flow would be established.

The calculations used to find the discharge rate are based on the method of Fauske (1965), as adapted by Cude (1975). Fauske worked on critical discharges of steam and water; in order to apply his method to other materials it is assumed that the ratio of the critical pressure at the throat to the upstream pressure for water systems (0.55) can be applied to other materials, provided the two phases are homogeneous and in mutual equilibrium. Thus:

$$P_c = 0.55\, P_1$$

The liquid fraction, $F_{vap}$, which flashes at $P_c$ is given by:

$$F_{vap} = \frac{C_{p1}\,(T_1 - T_c)}{H_{vap}}$$

where $T_c$ is the boiling point of the liquid at pressure $P_c$.

The mean density, $\rho_m$, of the two phase mixture is given by:

$$\rho_m = \frac{1}{\dfrac{F_{vap}}{\rho_g} + \dfrac{(1 - F_{vap})}{\rho_l}}$$

and the discharge rate, Q, is given by:

$$Q = C_d\, A_r \sqrt{2\,\rho_m\,(P_1 - P_c)}$$

The proportion of liquid entrained can be obtained by making assumptions similar to those recommended for Liquid Outflow in Section 4.1.1.

If $F_{vap}$ is greater than one (which is possible, but extremely unlikely for most materials), the method of calculation is not appropriate. The calculations can be repeated treating the release as a gas in which case the release rate will be underestimated. Alternatively, the ratio of $P_1$ can be adjusted to satisfy the condition that $F_{vap} = 1$, and the discharge calculations conducted using the appropriate thermophysical properties for this condition. On the other hand, if $F_{vap}$ is low it is simpler, and slightly conservative, to use the Liquid Outflow Model from Section 4.1.1.

---

| | |
|---|---|
| Outputs | This method provides a means for estimating the flashed fraction of a liquid discharge and thus enables the discharge rate from a two-phase flow to be calculated. |
| Inputs | (i) Discharge coefficient, $C_d$. For two-phase flow a value of 0.8 should be used.<br>(ii) Thermophysical data for the fluids, e.g. from DIPPR (1985)<br>(iii) Initial pressure of release $P_1$.<br>(iv) Selection of an effective open area, $A_r$, for the hole. |
| Assumptions | The method assumes that the two phases are homogeneous and in mutual equilibrium. |
| Accuracy | This is a simple empirical method which gives reasonably accurate results for the simple systems commonly of interest in hazard assessments. However, the accuracy of the method is questionable for discharges involving long lengths of pipeline where two-phase flow may develop within the line. |
| Application | This method may be applied to releases to atmosphere of saturated liquids stored under pressure at a temperature above the normal boiling point. |
| Resource Requirements | This calculation can be performed with a calculator. |
| Linkage with Other Models | The release rate of liquid is used to model pool behaviour; the release rate of gas and entrained aerosol is used to model jet or cloud behaviour. |

## 4.2  BEHAVIOUR IMMEDIATELY AFTER RELEASE

It is immediately after release that discharged materials show most variation in their behaviour since they are still strongly influenced by their origins, i.e. by the state of the inventory and the duration of the release. Unless ignited during this early stage, most releases, whatever their origins, develop into dispersed gas clouds. This section deals with the initial formation of a gas cloud for the following releases:

- a) A liquid release.
- b) A continuous jet.
- c) An instantaneous release.

### 4.2.1  Spreading Liquid Release

**METHOD**  Model for spreading and evaporating liquid.

**OUTPUT**  Spreading rate, giving pool diameter as a function of time after release; evaporation rate as a function of time.

**CONSTRAINTS**  The spreading model assumes a flat, level surface and a circular pool.

---

The Method

A release of a liquid with low volatility, stored at ambient conditions, does not usually represent a danger to people outside the plant because there will be little evaporation from the pool. It can affect people on-site if it ignites and it can affect the environment if it seeps into the soil and thence into water supplies, but it is the type of release that is most easily contained.

A release of a volatile liquid or of a cryogenic liquid, on the other hand, can represent a danger to people outside the plant because both releases will evaporate, and the gas could be carried off-site.

Immediately after release the liquid spreads out on the ground. It will spread until it meets an artificial boundary such as a dike or bund wall, until it reaches a minimum depth at which it no longer spreads, or until the evaporation rate is equal to the release rate so that the amount of liquid in the pool is no longer increasing. Also immediately after release, the liquid starts boiling off as it absorbs heat from the air, the ground or the sun. Mass is also lost from the pool when wind removes the evaporated material from the surface of the pool so that the material evaporates in order to restore the partial pressure. A volatile release will evaporate because of mass transfer by the wind only. A cryogenic release will evaporate because of both heat transfer and mass transfer. Both forms of evaporation continue after spreading has stopped, until all the liquid has evaporated.

The analyst needs to know the evaporation rate and duration in order to model the behaviour of the gas cloud. To calculate these the analyst must first consider the spreading process in order to find the radius of the pool as a function of time. This is essential to finding the evaporation rate as a function of time, because the heat transfer to the liquid, and mass transfer due to the wind depend on the surface area of the pool.

### The Spreading Process

The spreading model given here is that of Shaw and Briscoe (1978). This assumes a cylindrical pool spreading on a smooth level surface. The radius is given as follows:

**For an instantaneous spill:**

$$r = \left[\frac{t}{\beta}\right]^{\frac{1}{2}} \quad \text{where} \quad \beta = \left[\frac{\pi\rho_l}{8gm}\right]^{\frac{1}{2}}$$

**For a continuous spill:**

$$r = \left[\frac{t}{\beta}\right]^{\frac{3}{4}} \quad \text{where} \quad \beta = \left[\frac{\pi\rho_l}{32gW}\right]^{\frac{1}{3}}$$

Given the radius, the evaporation rate at a given time can be calculated.

| Surface | $\lambda_s$ (W/mK) | $a_s$ (m$^2$/s) |
|---|---|---|
| Concrete | 1.1 | $1.29 \times 10^{-7}$ |
| Ground (8% water) | 0.9 | $4.3 \times 10^{-7}$ |
| Dry sand | 0.3 | $2.3 \times 10^{-7}$ |
| Wet sandy ground | 0.6 | $3.3 \times 10^{-7}$ |
| Gravel | 2.5 | $11.0 \times 10^{-7}$ |

TABLE 4.1 : *Heat Transfer Properties of Some Surfaces*

### The Evaporation Process

Immediately after release most of the heat for boil-off is taken from the ground. The model given is taken from TNO (1979) which gives the evaporation rate, $m_g$, as:

$$ m_g = \frac{\lambda_s (T_a - T_b)}{H_{vap} (\pi a_s t)^{\frac{1}{2}}} $$

where $\lambda_s$ is the coefficient of heat conduction

$a_s$ is the thermal diffusivity.

Both are properties of the surface. Typical values are given in Table 4.1 above.

Once spreading has stopped, heat transfer from the ground decreases because the ground cools down. Eventually it becomes insignificant compared with mass transfer by the wind, which is independent of time and continues until all the liquid has evaporated. The model given for mass transfer by the wind is that of Sutton (1953) which gives the mass transfer rate $m_w$ as:

$$ m_w = a \left[ \frac{P_s M}{RT_a} \right] u^{\frac{(2-n)}{(2+n)}} r^{\frac{(4+n)}{(2+n)}} $$

where a and n are related to atmospheric stability as shown in Table 4.2 below.

| Stability Condition | n | a |
|---|---|---|
| Unstable | 0.2 | $3.846 \times 10^{-3}$ |
| Neutral | 0.25 | $4.685 \times 10^{-3}$ |
| Stable | 0.3 | $5.285 \times 10^{-3}$ |

TABLE 4.2: *Parameters in Sutton's Pool Evaporation Model*

### *Entrainment of Air into the Vapor*

The rate of entrainment of air into the vapor determines the density of the cloud and thus determines the choice of cloud dispersion model. The rate of entrainment will depend on such factors as windspeed, air stability, and the density of the vapor, and there is no accepted model. The simplest approach would be to choose a height $H_o$ over the pool, above which there is assumed to be no significant concentration of vapor. The mass evaporating will mix with the wind blowing through the volume between surface of the pool and $H_o$; the mass flow rate of air through this volume can be calculated and this gives the entrainment rate of air into the vapor.

### *Simplifications to the Model*

To apply the above equations properly the analyst will need to use a computer, and consider spreading and evaporation at different times after release; this computer analysis will give the evaporation rate at different times, for use in the subsequent gas dispersion calculations. An analyst without a computer will be unable to model the spreading and evaporation in detail, and will have to make some simplifying assumptions.

Some suggestions as to possible simplifying assumptions are given below. These simplification are based on finding the maximum pool diameter and using the evaporation rate due to wind for that diameter as the uniform, equivalent evaporation rate for the whole evaporation process. The evaporation rate due to wind is used instead of the evaporation rate due to heat transfer from the ground because this evaporation rate is independent of time and because it is usually reasonable to assume that by the time the pool has reached its maximum diameter the heat transfer from the ground will be negligible.

The duration of the vapor release can be approximated as follows:

$$\text{Duration} = \frac{\text{Mass of Liquid Released onto the Ground}}{\text{Uniform Evaporation Rate}}$$

The maximum pool diameter will depend on the topography of the ground near the release point, and on whether the release is continuous or instantaneous. Channels, drains, steps and walls can all influence the liquid spread. Often, for example, there is a spill containment provision, such as a low wall or dike (bund).

### For a continuous or instantaneous release with a dike

If there is a dike (or bund), the spreading will be cut short when the liquid reaches the dike so the maximum diameter of the pool will be the size of the dike.

### For a continuous release without a dike

Since there is no artificial boundary to stop the spread, the pool will only stop spreading when no more liquid is added to it. This will occur when the liquid inventory is exhausted, or when the evaporation rate equals the release rate of liquid onto the ground; at the moment when either of these occurs the pool will be at its maximum diameter.

The analyst should first calculate the radius, $r_w$, at which the evaporation rate from the wind equals the liquid release rate. The analyst should then calculate the time, $t_w$, at which the pool will reach this radius. If $t_w$ is less than the duration of the liquid release the maximum pool radius is given by $r_w$. If $t_w$ is greater than the duration of the liquid release the maximum pool radius is given by $r_d$, the radius corresponding to the duration of the liquid release.

### For an instantaneous release without a dike

This is a difficult case to model because, unlike the continuous release discussed above, it is not possible to say that spreading will stop when no more liquid is being added to the pool, since all of the liquid is released into the pool at the moment of release. On a flat, smooth, level surface such as that considered in this model, the pool will stop spreading only when it reaches some minumum thickness. On a completely smooth surface this minimum thickness will depend on surface tension; on a rough surface the minimum thickness will depend on the surface roughness. According to TNO (1979) there are no results linking minimum thickness to surface roughness. However, TNO recommends a lower limit of 5mm for very smooth surfaces; for rough surfaces the minimum thickness can be several centimetres; the value is left to the analyst's judgement. The maximum diameter of the pool can be calculated assuming that the liquid spreads instantaneously to the minimum thickness.

| Outputs | This method provides a way of estimating the spread of a non-reactive liquid. It gives the radius of a spill after a given time has elapsed, and the evaporation rate at that time. |
|---|---|
| Inputs | (i) Parameters in the equations above<br>(ii) Rate of liquid spillage, or mass of liquid released. |
| Assumptions | The spreading model makes the simplifying assumption that the initial release takes the form of a tall cylinder which then spreads under gravity. More complex models have been suggested but there is little justification for their use. The model also assumes that the ground is smooth and level, except for the possible prescence of a dike (bund) wall. |
| Accuracy | Accurate data on the behaviour of large spreading pools are sparse but suggest that these simple models are sufficiently accurate for use in hazard assessment. |
| Application | This method can be applied to the spreading and evaporation of volatile and cryogenic liquid releases. |
| Resource Requirements | A computer must be used in order to model the spreading and evaporation properly. However, if some simplifying assumptions are made a calculator can be used. |
| Linkage with Other Models | The evaporation rate is used in the appropriate cloud dispersion model. The pool size is used in the pool fire model. |

## 4.2.2          Jet Dispersion

**METHOD**       Simple Jet Model.

**OUTPUT**       For a given distance from the release point, the concentration and velocity on the jet axis, and the concentration profile at right angles to the axis.

**CONSTRAINTS**  Plume characteristics can be calculated where jet momentum is dominating the mixing process.

---

**The Method**   There are various models available for predicting the shape of a jet of material escaping under pressure; some of the models are simple, others are more complex. Complex models such as that of Emerson (1985) can involve large computational loads and require a computer. The model given here is the relatively simple one given in TNO (1979).

The TNO model applies to a jet of vapor at ambient conditions. In most cases the material will not be at ambient conditions immediately after outflow; the temperature will be lower than ambient temperature and, if the flow was choked, the pressure will be above ambient pressure. The jet is modelled as an equivalent jet with the same release rate as the real outflow, but with ambient outflow conditions.

This equivalent jet emerges from an orifice with an equivalent diameter $D_{eq}$, given by:

$$D_{eq} = D_o \sqrt{\frac{\rho_{go,a}}{\rho_{g,a}}}$$

where  $D_o$    is the diameter of the real orifice used in the outflow calculations.

$\rho_{go,a}$    is the relative density of the gas at outflow conditions, i.e. the density immediately after release, relative to air at ambient conditions.

$\rho_{g,a}$    is the density of the gas at ambient conditions, relative to air at the same conditions.

The flow is assumed to reach ambient conditions instantaneously, so that the equivalent orifice can be considered as being coincident with the real orifice.

### The Concentration Profile of the Jet

The concentration on the axis of the jet at a distance x from the orifice is given by:

$$c_m = \left[ \frac{\dfrac{b_1 + b_2}{b_1}}{0.32 \dfrac{x}{D_{eq}} \cdot \dfrac{\rho_{g,a}}{(\rho_{go,a})^{1/2}} + 1 - \rho_{g,a}} \right]$$

where $b_1$ and $b_2$ are distribution constants, given by:

$$b_1 = 50.5 + 48.2\rho_{g,a} - 9.95\rho_{g,a}^2$$

$$b_2 = 23.0 + 41.0\rho_{g,a}$$

The above equation can easily be rearranged to give x as a function of $c_m$; this can be used to calculate the length of the jet above a given concentration.

The concentration profile in a plane at right angles to the axis of the jet is given by:

$$\frac{c_{x,y}}{c_m} = e^{-b_2 (y/x)^2}$$

where $c_{x,y}$ is the concentration at a point a distance x from the orifice, and a distance y from the axis.

### The End of the Turbulent Mixing Phase

The speed of the jet will drop with distance along the axis, until, at a certain point on the axis, the jet speed will equal the wind speed. At this point the release has ceased to be a momentum jet, and must be modelled differently. The velocity distribution along the axis of the jet is given by:

$$\frac{u_m}{u_o} = \frac{\rho_{go,a}}{\rho_{g,a}} \cdot \frac{b_1}{4} \left[ 0.32 \frac{x}{D_{eq}} \cdot \frac{\rho_{g,a}}{\rho_{go,a}} + 1 - \rho_{g,a} \right] \cdot \left[ \frac{D_{eq}}{x} \right]^2$$

where $u_m$ is the velocity on the axis at a distance x from the orifice.

$u_o$ is the real outflow velocity of the release, calculated as follows:

$$u_o = \frac{\dot{m}_o}{C_d \rho_{go} \pi \left[ \frac{D_o}{2} \right]^2}$$

The analyst should calculate xw, the value of x at which um equals the wind speed, and the centre-line concentration cw corresponding to xw.

If cw is less than the minimum concentration of interest, e.g. the lower flammable limit, then the release need not be modelled further. The length of the jet with a concentration above the LFL can be obtained and used for calculating the size and effect of Jet Flames (see Section 4.4.2).

If cw is greater than the minimum concentration of interest, the release must then be modelled further using the appropriate plume dispersion model. If cw corresponds to a density which is equal to or less than the density of ambient air, then the Neutral, or Gaussian Dispersion Model should be used (Sections 4.3.2 and 4.3.3). If cw corresponds to a density greater than the density of ambient air, then a further test of density must be used, since the effective density of the plume as a whole might still be neutral or buoyant. The boundaries of the plume for the purposes of the transition to the dispersion models are not well-defined. If it is assumed that the boundaries occur where the concentration is 10% of the centre-line concentration, then 94% of the released material will be within the boundary. From this assumption the equivalent uniform density for the plume can be calculated. This is the density that should be used when choosing the dispersion model.

This model gives satisfactory results provided the concentrations of interest are above about 1 vol. %. It can therefore be used to find the dimensions of the LFL for most flammable gases. However, for most toxic gases the concentrations of interest are much lower than 1 vol. % so this model should not be used. To model the dispersion of a toxic gas the analyst should use either the dense plume model (Section 4.3.1), or a combination of the plume rise model (Section 4.3.3) and the neutral dispersion model (Section 4.3.2).

---

| | |
|---|---|
| **Outputs** | This method calculates the concentration and velocity on the axis of the jet at a given distance from the release point, and the concentration profile at right angles to the axis. |
| **Inputs** | (i) Discharge conditions of a sonic gas or a two-phase release.<br>(ii) Thermophysical properties of the material. |
| **Assumptions** | The model makes a number of simplifications and is therefore approximate. |
| **Accuracy** | This model is considered sufficiently accurate for most hazard analysis calculations. It is accurate down to concentrations of about 1%. It is really intended to model gas releases so, though it can be used to model two-phase releases, the accuracy decreases as the liquid fraction increases. |
| **Application** | This method can be used to estimate the behaviour and dispersion of a high velocity jet of vapour down to the LFL. Toxic releases may require different calculations. |
| **Resource Requirements** | This calculation can be performed using a calculator. |
| **Linkage with Other Models** | The dimensions and concentration of the jet are used in the appropriate cloud dispersion model and also in the jet fire model. |

## 4.2.3  Adiabatic Expansion

**METHOD**          Two-stage expansion model for instantaneous release of flashing liquid or pressurised vapour.

**OUTPUT**          Concentration and radius of expanded cloud.   Volume of the cloud and mass of air mixed into the cloud.

**CONSTRAINTS**     The model applies to rapid expansion with no heat exchange between the expanding mixed cloud and the surroundings.

---

**The Method**

After the instantaneous release of a flashing liquid or a pressurised gas, there will be a period of rapid expansion.   It is assumed that the process is so rapid that there is no time for heat exchange between the mixed cloud and the surroundings.  The expansion can therefore be treated as adiabatic.

The model given here is taken from TNO (1979).  The cloud is modelled as a hemisphere consisting of two zones: an inner zone or "core" of uniform concentration which contains 50% of the released mass, and a peripheral zone which has a Gaussian distribution of concentration.

This two-zone cloud is assumed to expand in two stages.   In the first stage, the gas or liquid aerosol expands down to atmospheric pressure.   In the process of this expansion the cloud acquires kinetic energy, known as "expansion energy".  In the second stage, the expansion energy carries the cloud further outward and also drives the turbulent mixing of air into the cloud.  The second stage is assumed to last until  the spreading velocity of the core falls below a given value

### Stage 1 : Expansion to Atmospheric Pressure

During the first stage the expanding gas or liquid does work against the atmosphere and some internal energy goes to increasing the kinetic energy of the substance.  If it is assumed that the increase in kinetic energy is given by $(P_1 - P_a)dV$, the initial expansion to atmospheric pressure can be treated as a reversible adiabatic process.   The energy of expansion becomes the difference between the initial and final energies, minus the work done on the atmosphere.   This first phase, idealised in this way, is isentropic.

For a **gas release**, the energy can be determined by considering a reversible adiabatic expansion from $P_1$ and $T_1$ with internal energy $U_1$ and volume $V_1$, to the state defined by $P_a$ and $T_2$ with internal energy $U_2$ and Volume $V_2$. The change in internal energy is then:

$$U_1 - U_2 = C_v ( T_1 - T_2 )$$

and the energy of expansion is:

$$E = C_v ( T_1 - T_2 ) - P_a ( V_2 - V_1 )$$

For a **liquid release**, the fraction, $F_{vap}$, that flashes is calculated by assuming that the entropy is constant during this idealised initial phase, i.e.

$$S_L(1) = (1 - F_{vap})S_L(2) + F_{vap}S_v(2)$$

where:

$$F_{vap} = \frac{S_L(1) - S_L(2)}{S_v(2) - S_L(2)}$$

$$= \frac{T_b [ S_L(1) - S_L(2) ]}{H_{vap}(2)}$$

As for a gas, the energy of expansion is the difference in initial and final internal energies minus the work done against the atmosphere:

$$E = \left\{ U_1(1) - \left[ ( 1 - F_{vap} ) U_L(2) + F_{vap} U_v(2) \right] \right\} - P_a ( V_2 - V_1 )$$

$$= H_L(1) - H_L(2) - ( P_1 - P_a ) V_1 - T_b \left[ S_L(1) - S_L(2) \right]$$

The differences in enthalpy and entropy are made consistent by using the heat capacity along the saturation curve to calculate both. For simplicity the heat capacity can be assumed to vary linearly between states (1) and (2).

### Stage 2 : Turbulent Mixing with Air

As a result of the impulse developed in the expansion, extensive turbulence is generated. This turbulence is the determining factor for further mixing of the gas cloud with the surrounding air. Once the expansion energy, E, has been obtained, the expression for the turbulent diffusion coefficient, $K_d$, is:

$$K_d = 0.0137 \, (V_{go})^{\frac{1}{3}} E^{\frac{1}{2}} \left[ \frac{(V_{go})^{\frac{1}{3}}}{t \, E^{\frac{1}{2}}} \right]^{\frac{1}{4}}$$

where $V_{go}$ is the volume of the gas at standard temperature and pressure.

The core radius, $r_c$, and the core concentration, $c_c$, as a function of time are given by:

$$r_c = 1.36 \left[ 4 K_d t \right]^{\frac{1}{2}}$$

$$c_c = \frac{0.0478 \, V_{go}}{\left[ 4 K_d t \right]^{\frac{3}{2}}}$$

As stated above, the second stage is considered complete when the spreading velocity of the core ($dr_c/dt$) falls below a given value. The choice of critical velocity is arbitrary but 1m/s is recommended since it is typical of fluctuating velocities in the natural atmosphere. If this velocity is combined with the above equations for E, $r_c$ and $c_c$, the following expressions are obtained for the core radius and concentration at the end of stage 2:

$$r_{ce} = 0.08837 \, E^{0.3} \, (V_{go})^{\frac{1}{3}}$$

$$c_{ce} = 172.95 \, E^{-0.9}$$

The dimensions of the peripheral zone of the cloud at the end of stage 2 are obtained by using the results of the Freon spill tests carried out by van Ulden (1974). In these tests it was found that at the end of expansion the peripheral cloud radius $r_{pe}$ was such that:

$$r_{pe} = 1.456 \, r_{ce}$$

### Mass of Air in the Cloud at the End of Expansion

The mass of air in the cloud must be calculated in order to calculate the density of the cloud; density is very important in the dispersion calculations given in Section 4.3.

For a **gas release** it is simple to calculate the mass of air in the cloud once the dimensions are known. The following formula is used:

$$V_{cloud} = \frac{M_{air}}{\rho_a} + \frac{M}{\rho_g}$$

For a **liquid release** in which liquid is still present at the end of expansion the calculations are more complicated and involve iteration.

In order to determine whether liquid is present the analyst should calculate $c_{ce}$; if $c_{ce}$ is greater than 1 then liquid is present. In calculating the vol./vol. concentration, $c_c$, the volume of the material is taken to be the volume if it were all vapour at the appropriate temperature and pressure; therefore $c_c$ can be greater than 1 if, and only if, liquid is present.

The amount of air mixed in with a cloud which still contains liquid is calculated iteratively by combining three equations.

The first equation expresses the heat balance between the elements of a cloud with a final temperature $T_3$ and a final vapour fraction $F_{vap3}$:

$$MC_{p1}T_1 + M_{air}C_{pa}T_a = M(1 - F_{vap3})C_{p3}T_3 + MF_{vap3}H_{vap} + M_{air}C_{pa}T_3$$

The second equation is a volume balance similar to that used above for gases:

$$V_{cloud} = \frac{M_{air}}{\rho_a} + \frac{F_{vap3}\,M}{\rho_g} + \frac{(1 - F_{vap3})\,M}{\rho_l}$$

The third equation states that the temperature and vapour fraction must be consistent with the requirement that the partial pressure should equal the saturated vapour pressure (s.v.p.) at $T_3$:

$$\text{s.v.p.} \; (T_3) = p_a \left[ \frac{F_{vap3} \, M}{\rho_g \, V_{cloud}} \right]$$

The mass of air should be adjusted until these three conditions are satisfied.

---

Outputs

The method gives the concentration and radius of the expanded cloud, the mass of air entrained in the cloud and the volume of the cloud.

Inputs

(i)   Initial volume, pressure and temperature of the released material
(ii)  Thermophysical properties of the material and of air

Assumptions

The model assumes an instantaneous release of material and rapid expansion in which no heat exchange takes place between the expanding mixed cloud and the surroundings.

Accuracy

The model is less accurate than the jet dispersion model because it is less well supported by experiments and observation. However, it is considered the best model currently available for large instantaneous releases and it is suitable for most hazard analyses.

Application

This method describes the initial behaviour of an instantaneous pressurised release. The output can be used for subsequent dispersion calculations.

Resource Requirements

The calculations can be performed with a calculator. However, for a liquid release in which liquid still remains at the end of expansion the calculations are not trivial because iteration is needed to find the mass of air mixed into the cloud.

Linkage with Other Models

The dimensions and density of the cloud at the end of expansion are used in the cloud dispersion models.

## 4.3 DISPERSION IN THE ATMOSPHERE

The dispersion of hazardous and pollutant materials in the atmosphere has been the subject of intense interest for some decades. This interest has resulted in the development of many different models for dispersion. The first models were developed in order to study the behaviour of pollutants discharged from vents and stacks. These pollutants usually form neutral plumes, i.e. plumes with densities similar to that of air; therefore the first models concentrated on neutral dispersion. More recently, the growth of interest in hazard analysis has been accompanied by a growth of interest in the behaviour of clouds with a density significantly different from that of air. In a hazard analysis the clouds which are denser than air are usually of most importance; clouds which are lighter than air will float upwards and are therefore likely to disperse harmlessly.

The dispersion of material in the atmosphere is a function of the stability of the air, the wind speed and the surface roughness, as described below:

a) **Air Stability**
   Stability is defined in terms of the vertical temperature gradient in the atmosphere. It is usually described using the system of categories developed by Pasquill. This system uses 6 (or sometimes 7) categories to cover unstable, neutral and stable conditions; the categories are ranges of stability identified by the letters A - F (or sometimes A - G). Neutral stability occurs typically when there is total cloud cover and is designated category D. Unstable conditions occur when the sun is shining because the warming of the ground increases convective turbulence; unstable conditions are designated by the letters A - C, with A as the least stable condition. Stable conditions occur on clear, calm nights when the air near the ground is stratified and free from turbulence, and are designated by the letters E and F; sometimes an additional category G is used for exceptionally stable conditions.

b) **Wind Speed and Surface Roughness**
   These factors are discussed together because they combine to influence local turbulence. The wind usually increases atmospheric turbulence and accelerates dispersion. The surface roughness of the ground induces turbulence in the wind which flows over it, and therefore affects dispersion.

All the factors discussed above appear in dispersion models. Some of the more recent and advanced models introduce more complex descriptions of turbulent mixing based upon eddy diffusivity. However, these models are so complex that they have not yet become widely used in hazard analysis; this section uses simpler models.

The models described in this section are:

a) **Dense Cloud Dispersion Model**                       Section 4.3.1
   This is a typical, simple "top-hat" model which describes the dispersion of the cloud while the cloud is spreading under gravitational forces. Once atmospheric turbulence has become the main cause of spreading the neutral dispersion model should be used.

b) **Neutral Cloud Dispersion Model**                     Section 4.3.2
   This is the Gaussian plume model which describes the behaviour of a neutral cloud.

c) **Plume Rise Model**                                   Section 4.3.3
   This is the Briggs plume-rise model whcih determines whether a plume will become air-borne.

## 4.3.1　　　Dense Cloud Dispersion

**METHOD**　　　Cox and Carpenter dense gas dispersion model.

**OUTPUT**　　　Gives cloud dimensions as a function of time for both steady continuous and instantaneous releases.

**CONSTRAINTS**　　Applicable only while the cloud is spreading due to gravitational forces.

---

**The Method**　　The model described here for dense cloud dispersion is that of Cox and Carpenter (1980); there are several other "top-hat" or "box" models in common use.

In these models an instantaneous release is represented as a cloud with a cylindrical or "top-hat" shape, which slumps under gravity so that it spreads radially relative to its centre; at the same time the cloud is moving with the wind so that the centre is moving at the windspeed. The shape and movement of an instantaneous cloud with radius R and height H are shown in Figure 4.1.

A steady continuous release is represented as a plume with a centre-line parallel to the wind direction. In the cross-wind direction the plume is assumed to have a rectangular cross-section, with width 2L and height H. The width and height vary along the length of the plume, increasing as the distance from the source increases, as shown in Figure 4.2. This figure shows three cross-sections through the plume at different distances from the source. Since the releases rate is assumed constant, these cross-sections could be regarded as representing either the dimensions of a single element of the plume as three different points in time, or the dimensions of three different elements in the plume at the same point in time. All the elements in the plume experience the same dispersion process so that, at a given distance from the source, the dimensions, concentration, density and temperature of the plume will be constant throughout the release. This means that in order to model the dispersion of a steady continuous plume it is necessary to consider only the dispersion of one segment of the plume as it progresses downwind.

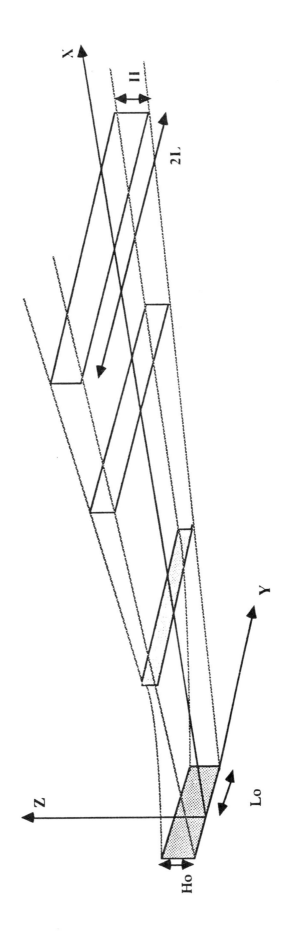

FIGURE 4.1 : *Dimensions and Movement of a Continuous Dense Cloud*

FIGURE 4.2: *Dimensions of Instantaneous Dense Cloud*

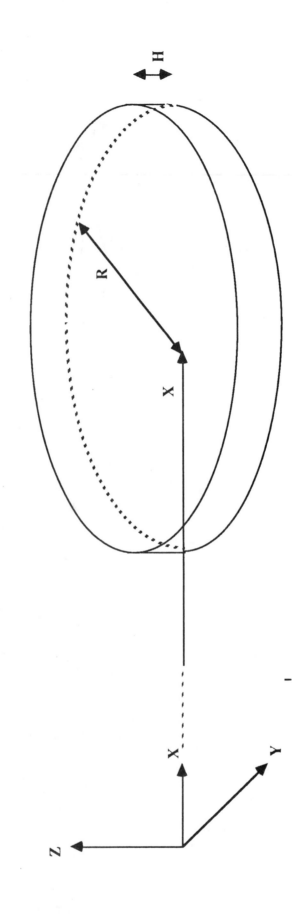

## Initial Dimensions of the Cloud

For an **instantaneous** release, the initial amount of air mixed into the cloud is calculated using the methods given in Section 4.2.3. The mixed cloud is taken to be a cylinder of uniform density, with a chosen ratio of initial height $H_o$ to initial radius $R_o$ and the same volume as the hemispherical cloud of Section 4.2.3. The height-to-radius ratio is usually taken as 1.

For a **continuous release from a pressurised source**, the initial amount of air in the cloud and the dimensions of the cloud are obtained from the jet dispersion model in Section 4.2.2. In this model, at the point where turbulent mixing stops, the cross sectional boundary of the cloud is assumed to occur where the concentration is 10% of the centre-line concentration. the plume is therefore assumed to have a circular cross section at the end of turbulent mixing. In the dense model, the cloud is modelled as a plume with a square cross section of the same area as the circular cross section as the jet model.

For a **continuous release from an evaporating pool**, the initial density of the cloud is obtained by calculating the amount of air entrained in the evaporating vapour. The initial semi-width, $L_o$, can be taken as the radius of the pool. The initial height, $H_o$, is the height considered when calculating the mass flow rate of wind across the pool (as described in Section 4.2.1)

## Spreading of the Cloud

For all releases, lateral spreading due to gravity is given by:

$$\frac{dR}{dt} \text{ or } \frac{dL}{dt} = \sqrt{kgH(\rho_{c,a} - 1)}$$

where the constant k is given various values by different authors; inviscid flow theory suggests that k should be 2 but tests on real flows such as those of van Ulden (1974) give it a value of 1 which is the value now used.

## Mixing of Air into the Cloud

For both continuous and instantaneous releases, air mixes into the cloud at two surfaces: at the edges of the cloud, where air mixes into the cloud as it spreads; and at the top of the cloud, where atmospheric turbulence and the difference in densities determine the amount of mixing. Combining the entrainment rates for the top and edge of the cloud gives the total entrainment rate.

For edge mixing, the entrainment rate $Q_e$ (in $m^3/s$) is given by:

$$Q_e = \gamma \frac{dR}{dt}. \text{ (area of edge surface of the cloud)}$$

where $\gamma$ is the dense cloud edge-mixing coefficient. It indicates the proportion of the air in front of the advancing cloud that mixes in with the cloud instead of being displaced upwards. Different values have been suggested for this coefficient; Cox and Carpenter use 0.6.

For mixing through the top surface of the cloud, the entrainment rate (in $m^3/s$) is obtained by multiplying the entrainment velocity $U_e$ (in $m/s$) by the area of the top surface of the cloud. $U_e$ is given by:

$$U_e = \frac{\alpha u_1}{Ri}$$

where $Ri$ is the Richardson number, given by:

$$Ri = \frac{gl(\rho_{c,a} - 1)}{(u_1)^2}$$

The expressions for $U_e$ and $Ri$ involve several terms which require further explanation and some suggestions as to suitable values:

$\alpha$    is an empirical constant which should be taken as 0.1

$u_1$    is the longitudinal turbulence velocity. $u_1$ is given by:

$$u_1 = \left[ \frac{u_1}{u^*} \right] \cdot \left[ \frac{u^*}{u} \right] \cdot u$$

where $u$ is the wind-speed and $u^*$ is the friction velocity.
$(u_1/u^*)$ is dependent primarily on stability. Monin (1962) gives the following approximate values:

| Stability | | $u_1/u^*$ |
|---|---|---|
| Very unstable | (A/B) | 3.0 |
| Neutral | (C/D) | 2.4 |
| Very stable | (E/F/G) | 1.6 |

$(u^*/u)$ is dependent primarily on surface roughness, as shown by Sutton (1953). For open terrain a typical value is 0.1.

$l$    is the turbulence length scale in metres. $l$ depends on atmospheric stability and on the cloud height, H. Some values for $l$ are given in Table 4.3 below, which is based on data presented by England et al. (1978)

| H | Stability Category | | | | | | |
|---|---|---|---|---|---|---|---|
| (m) | A | B | C | D | E | F | G |
| 10 | 18.0 | 15.0 | 12.0 | 10.0 | 8.0 | 7.0 | 6.0 |
| 20 | 30.0 | 25.0 | 21.0 | 18.0 | 16.0 | 14.0 | 12.0 |
| 30 | 41.0 | 34.0 | 29.0 | 25.0 | 22.0 | 20.0 | 17.0 |
| 50 | 62.0 | 52.0 | 44.0 | 39.0 | 35.0 | 31.0 | 27.0 |
| 75 | 84.0 | 71.0 | 60.0 | 52.0 | 48.0 | 43.0 | 37.0 |
| 100 | 105.0 | 85.0 | 74.0 | 64.0 | 60.0 | 54.0 | 46.0 |

TABLE 4.3 : *Turbulence Length Scale as a Function of Cloud Height H and Stability*

### Thermodynamics of Mixing

As warm air mixes into the cloud it adds its heat to the cloud, either increasing the temperature of the cloud, or vaporising some of the liquid in the cloud (if any is left). Heat is also transferred to the cloud from the ground; this heat transfer can be considerable, since the temperature difference between the cloud and the ground can be large, and the dispersion process can last a long time. The heat transfer from the ground can be due to either forced or natural convection. Both heat transfer rates should be calculated, and the larger value used to calculate the total heat transfer to the cloud, and thence the new temperature or liquid fraction of the cloud. This will give the new volume of the cloud.

The **natural convection** correlation is taken from McAdams (1954), simplified to give the heat flux $q_n$ (in J/m$^2$s) as:

$$q_n = h_n ( T_c - T_g )^{\frac{4}{3}}$$

A typical value for the coefficient $h_n$ is 2.7 J/m$^2$.s.K$^{4/3}$ for a cloud of LNG vapour.

The total heat transfer from natural convection is obtained by multiplying $q_n$ by the surface area of the bottom of the cloud.

The local heat flux $q_f$ from **forced convection** is given ( in J/m$^2$s) by:

$$q_f = \frac{C_f \rho_c C_p u' ( T_c - T_g )}{2}$$

where $C_f$ is the friction factor, given by $2 (u*/u)^2$.

u' is the vector sum of the wind and slumping velocities.

The total heat transfer rate from forced convection is obtained by integrating $q_f$ over the surface of the cloud.

### Transition to Neutral Cloud Dispersion

This dense cloud model can be used to model cloud spreading until the lateral spreading rate caused by atmospheric turbulence exceeds the spreading rate caused by gravity; once this has happened the cloud behaviour must be modelled using the neutral dispersion model given in Section 4.3.2. This transition in spreading rate occurs when :

$$\frac{dR}{dt} \ \text{or} \ \frac{dL}{dt} = \frac{d\sigma_y}{dt} = \frac{d\sigma_y}{dx} \cdot u$$

where $\sigma_y$ is the dispersion parameter in the cross-wind direction. It describes the increase in the cloud radius or cross-wind width as the cloud drifts down-wind; it is dependent primarily on stability. A fuller description of this parameter and its dependence on time is given in Section 4.3.2.

### Summary of the Dense Cloud Model

a) **Need for a computer**
   The dense cloud model is the most complex model in the manual. It is not possible to arrange the above equations to give a single, simple equation giving, say, the cloud radius as a function of time; the dispersion has to be modelled by considering the behaviour of the cloud at different points in time. It might be possible to do this using a calculator but it would be extremely laborious, so it is for this model that a computer would be particularly useful.

b) **Suggested procedure for calculations**
   At each step in the modelling the analyst must consider spreading, mixing, and heat transfer. These processes have been discussed separately above and their interaction is discussed below. This interaction is considered by taking one time-step in the modelling, and suggesting the order and method in which the processes should be calculated for this time step:

   1. *Calculate spreading rate given dimensions from previous time-step*
      If this rate is assumed constant throughtout the time-step, the new dimensions of the cloud, $H_s$ and $R_s$ (or $L_s$), can be calculated, assuming the volume of the cloud is constant during the time-step. These new dimensions are used in calculating the air-mixing and convection.

   2. *Calculate the mixing of air into the cloud*
      This is calculated from the current values of $dR/dt$, $H_s$ and $R_s$.

4.  *Calculate new density and concentration of cloud.*
    If the cloud contains vapor only, the heat added by the warm air and transferred from the ground will be used to raise the temperature of the cloud. If the cloud contains liquid the heat will be used to vaporise some of the liquid. Using heat and mass balances the new density and concentration can be calculated.

5.  *Calculate new dimensions of cloud*
    From the new density and volume, the new height $H_n$ can be calculated. The width of the cloud is set by the slumping process and is not affected by the mixing or heating.

6.  *Proceed to the next time-step*
    The dimensions and density at the end of the time step are used as initial conditions in the next time step.

---

| | |
|---|---|
| Outputs | This method gives the spreading and downwind displacement of the cloud. |
| Inputs | i)   Initial cloud dimensions and composition.<br>ii)  Wind speed and atmospheric stability.<br>iii) Thermophysical data for gas and air.<br>iv) Surface roughness of the ground. |
| Accuracy | Many comparisons and some tests have been conducted to validate these models. Usually, they can be calibrated to fit the test data. These methods are regarded as sufficiently accurate for modelling a gas cloud in a relatively flat environment with no significant obstruction. If these conditions are not met the results of this model should be regarded with caution. Unfortunately, models which deal with rough surfaces or obstructions are complex and unproven, and therefore outside the scope of this manual. |
| Application | These methods can be used to examine the behavior of clouds which are heavier than air. |
| Resource Requirements | These calculations must be performed using a computer. |
| Linkage with Other Models | The dimensions and properties of the cloud are used in the neutral gas dispersion model; they are also used in calculating toxic effects and the effects of flash fires and explosions. |

## 4.3.2      Dispersion of a Neutral-Density Cloud

**METHOD**      Gaussian model for neutral clouds.

**OUTPUT**      Cloud correlations for both instantaneous and continuous releases.

**CONSTRAINTS**      Applicable only to those phases of gas dispersion dominated by atmospheric turbulence.

---

**The Method**

This neutral dispersion model should be used for clouds in which the spreading rate due to turbulence is greater than the spreading rate due to gravity. This behavior will be found in clouds whose density immediately after release is neutral; such behavior is also found in initially-dense clouds whose density has been reduced by atmospheric mixing and heat transfer from the ground.

The concentration profile for these clouds is assumed to be Gaussian. Since most of the models which are used to model the behaviour of the release before neutral dispersion starts do not assume a Gaussian concentration profile some assumptions must be made in order to match the previous model to the Gaussian model. The assumptions for the various models are discussed below:

### Dense Cloud Model

The centre-line concentration, $c_m$, and cross-wind width, $w_m$, at the matching point are assumed to be the same for the dense and Gaussian models. In addition, in order to model the cloud as a neutral cloud, a "virtual source" for the cloud at the matching point must be calculated. This is necessary because the equations for the concentration of the cloud at the matching point all involve the distance travelled from the source; the true source cannot be used as the source in the Gaussian model because the cloud travelled some of the distance from the true source as a dense cloud so its concentration does not obey the neutral-cloud equations. Instead, a virtual source is used; the cloud is considered as starting at a point a distance $x_v$ upstream of the matching point, where it had a crosswind width $w_v$. $x_v$ can be found by trying different values in the appropriate equation for centre-line concentration (given later in the section) until the correct concentration, $c_m$, is obtained. $w_v$ is found by using the following:

$$w_m = w_v + 2\sigma_{ym}$$

where $\sigma_{ym}$ is the cross-wind dispersion parameter at the matching point. It is a function of atmospheric stability and $x_v$. There are dispersion parameters for 3 different directions:

       downwind ($\sigma_x$);

       crosswind ($\sigma_y$);

       vertical ($\sigma_z$).

They describe the increase in the cloud dimension in the appropriate direction as the cloud drifts down-wind. They are extremely important in neutral dispersion and will be discussed in more detail later in the section.

## *Jet Dispersion Model*

The jet dispersion model does assume a Gaussian concentration profile. However, the rate of dilution in jet dispersion is greater than the rate in neutral dispersion so the analyst cannot simply use the true release rate and source in the neutral dispersion calculations because this will give concentrations that are too high. Instead the analyst must use an upwind virtual source with the same release rate as the real release rate. The location of this virtual source is found by matching the centre-line concentration of the end of the virtual cloud with the concentration at the centre of the jet. The cross wind width of the plume at the matching point is the diameter of the plume which includes all concentrations down to 10% of the concentration at the centre-line (as discussed in Section 4.2.2).

## *Neutral Instantaneous Releases*

If such a release is pressurised it will undergo adiabatic axpansion in which it is modelled as a two-zone hemisphere (Section 4.2.3). The outer radius at the end of expansion can be calculated and used as the initial cross-wind width in the Gaussian model. The adiabatic expansion is very rapid so the cloud will not have time to travel far; therefore, the true source location can be used as the source in the neutral dispersion calculation.

## *Neutral Evaporation from Liquid Pools*

The vapour produced by liquid pools is assumed to be of uniform density. The matching is done in the same way as for dense clouds, i.e. the centre-line concentration and cross-wind width are matched. The procedure is as described above.

Once the analyst has determined the location of the source, whether true or virtual, and the dimensions of the cloud at the source, the concentration of the cloud as it moves with the wind at a speed u can be calculated. The calculations are different for instantaneous and continuous releases. The equations obtained by Pasquill (1961) for instantaneous and continuous releases are given below, followed by a discussion of dispersion parameters, including some formulae for calculating these parameters. It should be noted that in the following equations the absorption at the ground is assumed to be zero, so that the cloud is "reflected" at the ground. This can be a conservative assumption for some materials.

### Instantaneous Release

For a release of a mass Q* at ground level, the concentration at a time t after the cloud has left the source is given by:

$$c(x,y,z,t) = \frac{2Q^*}{(2\pi)^{\frac{3}{2}} \sigma_x \sigma_y \sigma_z} \cdot \exp\left[ -\frac{1}{2} \left\{ \frac{(x-ut)^2}{\sigma_x^2} + \frac{y^2}{\sigma_y^2} + \frac{z^2}{\sigma_z^2} \right\} \right]$$

where x, y, and z are measured from the source point.

For ground-level concentrations at the cloud centre-line the above equation reduces to:

$$c(x,0,0,t) = \frac{2Q^*}{(2\pi)^{\frac{3}{2}} \sigma_x \sigma_y \sigma_z} \cdot \exp\left[ -\frac{1}{2} \left\{ \frac{(x-ut)^2}{\sigma_x^2} \right\} \right]$$

### Continuous Release

For a release rate of Q at ground level the concentration is given by:

$$c(x,y,z) = \frac{Q}{\pi \sigma_y \sigma_z u} \cdot \exp\left[ -\frac{1}{2} \left\{ \frac{y^2}{\sigma_y^2} + \frac{z^2}{\sigma_z^2} \right\} \right]$$

where x, y and z are measured from the source point.

For ground-level concentration at the cloud centre-line the above equation reduces to:

$$c(x,0,0) = \frac{Q}{\pi \sigma_y \sigma_z u}$$

| Stability Category | | Parameter | | | |
|---|---|---|---|---|---|
| | | a | b | c | d |
| Very unstable | A | 0.527 | 0.865 | 0.28 | 0.90 |
| Unstable | B | 0.371 | 0.866 | 0.23 | 0.85 |
| Slightly unstable | C | 0.209 | 0.897 | 0.22 | 0.80 |
| Neutral | D | 0.128 | 0.905 | 0.20 | 0.76 |
| Stable | E | 0.098 | 0.902 | 0.15 | 0.73 |
| Very stable | F | 0.065 | 0.902 | 0.12 | 0.67 |

TABLE 4.4 : *Terms in Expressions for Dispersion Parameters, as Functions of Atmospheric Stability*

### Dispersion Parameters

As mentioned above, these parameters describe the increase in the dimensions of the cloud as the cloud drifts down-wind. There are several different formulae for these parameters. The ones given here are relatively simple ones taken from TNO (1979). They are:

$$\sigma_y = ax^b$$

$$\sigma_z = cx^d$$

where x is the distance down-wind of the source.

No formula is given for $\sigma x$. For instantaneous releases it is assumed that:

$$\sigma_x = \sigma_y$$

For continuous releases the down-wind dimension of the cloud is assumed to be negligible. Table 4.4 above gives the values for the parameters a, b, c and d suggested by TNO for different stability categories; these values are valid where x is greater than 100m.

| Outputs | The model gives cloud concentrations at distance down-wind, for both instantaneous and continuous releases. |
|---|---|
| Inputs | (i)  Released mass, or release rate |
| | (ii)  Dimensions and concentration of the cloud at the matching point |
| | (iii)  Windspeed and atmospheric stability |
| Assumptions | The dispersion is based on the assumption of Gaussian distributions of turbulence in the atmospheric boundary layer. |
| Accuracy | If the correct wind speed and stability category are used, this method gives concentrations that agree with experimental results to within a factor of 2. This should be satisfactory for the purposes of a hazard assessment. However, the equations are applicable only for down-wind distances greater than 100m and less than 100km; fortunately, this restriction should not be a great handicap to most hazard analyses. |
| Application | The method determines the concentration distribution of clouds of neutral density. |
| Resource Requirements | The calculations can be performed with a calculator; however, use of a computer will enable the analyst to consider more points in time. |
| Linkage with Other Models | The cloud dimensions and concentrations are used for calculating toxic effects and the effects of flash fires and explosions. |

## 4.3.3    Dispersion of a Buoyant Plume

**METHOD**    Briggs' Buoyant Surface Release Model to determine whether a plume on the ground will become air-borne, followed by Briggs' Plume Rise Model to calculate the trajectory of the release, and Gaussian dispersion to calculate the ground level concentration.

**OUTPUT**    Plume trajectory and concentrations.

**CONSTRAINTS**    Based upon empirical observations.

---

**The Method**

There are several stages involved in modelling a plume of material with a density less than that of the surrounding air. First, the analyst must calculate whether the release will lift off the ground, or whether the downward effects of turbulence will keep the plume on the ground. This calculation is only necessary for a release at ground level; an elevated release can be assumed to rise, since most elevated releases are from engineered release points such as stacks, and if the stack is properly designed, there should be no downward force. If the plume does not lift off, it should be modelled using the Gaussian dispersion model in Section 4.3.2. If the plume does lift off, the next stage in the modelling is to calculate the rise of the plume as a function of distance down-wind. This final stage is to calculate the ground-level concentration at different distances down-wind, using the Gaussian dispersion model with the appropriate values for plume elevation.

If plume rise is not taken into account, the analysis will ignore the possibility that an air-borne release could ignite in mid-air; this has happened on several occasions with air-borne releases of hydrogen. In addition, the ground level concentrations calculated will be too high since the centre-line concentrations will be taken as the ground level concentration.

### Stage 1 : Lift-Off from the Ground

For a surface release of buoyant material, it is not necessarily clear that the material will lift off the surface under the action of buoyancy forces. The downward effects of turbulence might be dominant; these can be intense near the ground due to friction effects and obstacles. For the cloud to lift off, the lower parts of the cloud edges must move inwards due to the external hydrostatic pressure acting against the spreading influence of the turbulent dispersion. Briggs (1976) has suggested that a criterion may be developed by comparing a characteristic lateral turbulent spreading velocity with a characteristic inward movement associated with buoyancy. This arises from considerations of the drawing-in of the cloud sides near the surface as the bulk of the cloud starts to rise.

For the velocity $U_i$ of the inward movement, Briggs uses:

$$U_i = \left[ \, gH \, ( \, 1 - \rho_{c,a} \, ) \, \right]^{\frac{1}{2}}$$

For the turbulent spreading velocity, Briggs uses the friction velocity $U^*$. The friction velocity is discussed in Section 4.3.1 on dense cloud dispersion. In order to predict lift-off Briggs therefore defines a parameter, $L_P$, given by:

$$L_p = \left[ \, \frac{U_i}{U^*} \, \right]^2$$

and he finds that lift-off occurs when this is greater than about 2.5 for instantaneous releases of roughly hemispherical shape and for continuous releases of roughly semi-cylindrical shape at the ground. It should be noted, however, that in other cases the critical value could be different, but in the absence of other information to the contrary, this value of 2.5 is generally adopted.

### Stage 2 : Rise of the Plume

The rise of the plume as a function of distance down-wind depends on such factors as the relative density of the plume, the windspeed, and the atmospheric stability. The stability in particular has a very important effect on the final height reached by the plume, and the concentrations experienced at ground level. In neutral conditions the plume will rise steadily until it disperses. In stable conditions the plume rises until it encounters warm air; it then no longer experiences a buoyancy force so no longer rises. In unstable conditions the cloud rises more quickly than in the neutral condition, but strong eddy currents may carry material back to ground level; this is therefore a potentially hazardous condition, but it is also very difficult to describe. The models given here apply only to stable and neutral conditions.

The models used for plume rise are those of Briggs (1969). These models were developed from studies of plumes from stacks so they contain terms which are more applicable to these releases to surface releases; however, analogies can be found for these terms to allow the models to be used for ground releases.

For all stability conditions the initial rise of the cloud is as follows:

$$\Delta h = 1.6 \frac{x^{\frac{2}{3}}}{u} \cdot F^{\frac{1}{3}}$$

where F is a buoyancy flux parameter given by:

$$F = ( 1 - \rho_{go,a} ) \, g \, v_o w_o^2$$

For neutral conditions the plume will continue to rise indefinitely, although the above equation is valid for only a part of the rise. The equation which describes the next part of the plume rise is complex and cumbersome; Briggs suggests the conservative approximation that the plume rises according to the equation above until it reaches a ceiling height at a distance downwind given by:

$$x = 6.49 \, F^{\frac{2}{5}} h_s^{\frac{3}{5}} \qquad\qquad ( h_s < 300m )$$

$$x = 201.9 \, F^{\frac{2}{5}} \qquad\qquad ( h_s > 300m )$$

where hs is the stack height. For a release from the ground a nominal stack height such as 1m should be used.

For stable conditions the plume rise equation given above holds approximately to a downwind distance of:

$$x = 2.4 \frac{u}{\sqrt{s}}$$

where s is a stability parameter given by:

$$s = \frac{g}{T} \cdot \frac{\partial \theta}{\partial z}$$

$\partial\theta/\partial z$ is a measure of the atmospheric temperature gradient. For stable conditions $\partial\theta/\partial z$ is in the range of 0.01-0.03 °C/m.

Beyond this down-wind distance the plume levels off at about:

$$\Delta h = 2.9 \left[ \frac{F}{us} \right]^{\frac{1}{3}}$$

For unstable conditions the equation for initial plume rise could be used but it could give non-conservative results.

## Stage 3 : Dispersion of Buoyant Plume

The dispersion is modelled using the Gaussian model described in Section 4.3.2. The ground-level concentration when the plume is at a distance x down-wind, and a height h above the ground is given by:

$$c(x,y,h) = \frac{Q}{\pi \sigma_y \sigma_z u} \cdot \exp\left[-\frac{1}{2}\left\{\frac{y^2}{\sigma_y^2} + \frac{h^2}{\sigma_z^2}\right\}\right]$$

---

**Outputs**
The model gives the plume trajectory and the ground-level concentration.

**Inputs**
(i)    Released mass, or release rate
(ii)   Dimensions and concentration of the cloud at the matching point
(iii)  Windspeed and atmospheric stability

**Assumptions**
The dispersion is based on the assumption of Gaussian distributions of turbulence in the atmospheric boundary layer.

**Accuracy**
If the correct wind speed and stability category are used, this method gives concentrations that agree with experimental results to within a factor of 2. This should be satisfactory for the purposes of a hazard assessment. However, the equations are applicable only for down-wind distances greater than 100m and less than 100km; fortunately, this restriction should not be a great handicap to most hazard analyses.

**Application**
The method determines the trajectory and concentration distribution of buoyant plumes.

**Resource Requirements**
The calculations can be performed with a calculator; however, use of a computer will enable the analyst to consider more points in time.

**Linkage with Other Models**
The trajectory and concentrations of the plume are used to calculate toxic effects and the effects of flash fires and explosions.

*Consequence Calculations*

## 4.4 FIRES

Fires can be categorised as follows:

-   **Pool fire  (Section 4.4.1)**
    e.g. a tank fire or a fire from a pool of fuel spread over the ground or water.

-   **Jet fire (Section 4.4.2)**
    from the ignition of a jet of flammable material.

-   **Fireball and BLEVE (Section 4.4.3)**
    (Boiling Liquid Expanding Vapour Explosion) resulting from the overheating of a pressurised vessel by a primary fire. This overheating raises the internal pressure and weakens the vessel shell, until it bursts open and releases its contents as a large and very intense fireball.

-   **Flash Fire (Section 4.4.4)**
    involving the delayed ignition of a dispersed vapour cloud which does not cause blast damage.  That is, the flame speed is not as high as in an unconfined vapour cloud explosion (Section 4.5), but the fire spreads quickly throughout the flammable zone of the cloud.

A fire affects its surroundings primarily through the radiated heat which is emitted.  If the level of heat radiation is sufficiently high, other objects which are flammable can be ignited.  In addition, living organisms may be burned by heat radiation.  The damage caused by heat radiation can be calculated from the dose of radiation received; a measure of the received dose is the energy per unit area of the surface exposed to the radiation over the duration of the exposure.  Alternatively, the likely effect of radiation may be estimated by using the power per unit area received.

The radiation effect of a fire is normally limited to the area close to the source of the release (say within 200m).  In many cases, this means that neighbouring communities are not affected.  However, there are some types of fire which could have a more extensive effect.

Simplified models for calculating the radiation from pool fires, jet fires, BLEVEs and flash fires are given in Sections 4.4.1 to 4.4.4.  The damage levels associated with different levels of radiation are discussed in Section 4.4.5.  A summary of some potential sources of ignition is given in Appendix III.

### 4.4.1      Pool Fires

**METHOD**  Use of classical empirical equations to determine burning rates, heat radiation and incident heat.

**OUTPUT**  Heat intensity and thereby an indication of the potential to cause damage or casualties.

**CONSTRAINTS**  This empirical model for calculating the form of the fire has been validated only for relatively small fires.

---

**The Method**

The model employed in the estimation of pool fires is derived from TNO (1979) and involves the use of classical empirical equations to determine burning rates, heat radiation and incident heat.

For a liquid with a boiling point above ambient temperature, the rate of burning of the liquid surface per unit area is given by:

$$\frac{dm}{dt} = \frac{0.001 \, H_c}{C_p (T_b - T_a) + H_{vap}}$$

For a liquid with a boiling point below ambient temperature, the expression is:

$$\frac{dm}{dt} = \frac{0.001 \, H_c}{H_{vap}}$$

The total heat flux from a pool of radius r is given by:

$$Q = \frac{(\pi r^2 + 2\pi rH) \cdot \left[\dfrac{dm}{dt}\right] \cdot \eta \, H_c}{72 \left[\dfrac{dm}{dt}\right]^{0.61} + 1}$$

where $\eta$ is an efficiency factor which normally takes a value in the range 0.13 to 0.35; in the absence of better data, the conservative value 0.35 should be used.

H is the flame height calculated as:

$$H = 84 \, r \left[\frac{(dm/dt)}{\rho_a (2gr)^{1/2}}\right]^{0.6}$$

When the heat flux at the surface of the pool fire has been calculated, the heat incident upon nearby objects may then be determined. A simplified method assumes that all the heat is radiated from a small vertical surface at the centre of the pool. For a ground pool, the heat incidence at a distance R from the pool centre is given by:

$$I = \frac{TQ}{4 \pi R^2}$$

where T is the transmissivity of the air path. In the absence of good data to the contrary, the transmissivity is set to unity; this gives conservative results.

---

| | |
|---|---|
| Outputs | The method provides estimates of radiative flux from a pool fire. |
| Inputs | (i) Thermophysical properties for the materials of interest<br>(ii) Fraction of heat liberated as radiation<br>(iii) Transmissivity of the air path to a receiver. |
| Assumptions | These methods are based upon empirical correlations. |
| Accuracy | Generally considered valid for the types of pool fire which might affect people and property away from the site. |
| Application | This method may be applied to the estimation of effects of pools of fuel which ignite, including tank fires and spreading pools on land. |
| Resource Requirements | These calculations can be performed with a calculator. |
| Linkage to Other Models | The heat flux from the fire is used to calculate fire damage. |

## 4.4.2 Jet Fires

**METHOD**      Use of radiation formula to determine intensity of radiated heat.

**OUTPUT**      Heat intensities.

**CONSTRAINTS**  Estimated radiation levels close to the base of the flame may be subject to error due to flame lift-off at the source.

---

**The Method**

The model employed in estimating thermal radiation effects from jet flames is an extension of the model used for jet dispersion including wind effects (API RP521,"Flare Radiation"). The flame is modelled as a series of point sources spaced along the centre line of the jet with all sources radiating equal quantities of heat $Q_p$.

The radiated heat, $Q_p$, for a release rate Q and $n_p$ point sources is given by:

$$Q_p = \eta \, Q \, H_c$$

where $\eta$ is an efficiency factor conservatively taken as 0.35.

The size of the jet is obtained from the jet dispersion model in Section 4.2.2. The most logical choice for jet length would be the length of the axis from the orifice to the LFL. However, in terms of heat flux at the flame surface a better match is obtained by including concentrations down to LFL/1.5 in the flame. The choice of $n_p$ is rather arbitrary; 5 points are usually adequate for a hazard analysis.

The radiation, I, from a particular point in the flame to a receptor at distance, r, is given by:

$$I = \frac{X_g Q_p}{4 \pi r^2}$$

where $X_g$ is an emissivity factor which depends on the material being burnt; for jet fires $X_g$ can be taken as 0.2.

The total heat flux at a distance from the flame is the summation of the radiation from each point in the flame.

It should be noted that this model does not take into account the effect of wind on the flame shape. Wind has little effect on a jet from a high-pressure source since the velocity of the jet is much greater than that of the wind. However, the shape of a jet from a low-pressure source, such as a controlled vent, can be changed significantly by the wind; the effect of the wind will be to increase the incident radiation for down-wind receptors.

| | |
|---|---|
| Outputs | The calculations produce an estimate of the radiative heat flux which is received at a position near the flame. |
| Inputs | (i) Fuel input rate<br>(ii) Length of jet<br>(iii) Location of receptor |
| Assumptions | It is generally assumed for convenience that the flame will have approximately the same length as an unignited jet. If detailed radiation envelopes are required, the iterative procedures necessary to accomplish the calculations are best conducted by computer. |
| Accuracy | The method is not accurate at the base of a flame if lift-off of the flame occurs, or if the jet is bent over by the wind. |
| Application | These simplified methods may be used to estimate radiation levels from jet releases of flammable material. |
| Resource Requirements | The calculations can be performed with a calculator. |
| Linkage with Other Models | The heat flux from the flame is used to calculate fire damage. |

### 4.4.3 Fireballs and BLEVEs

**METHOD**     Empirical correlation of fireball radius, duration, and radiation output.

**OUTPUT**     Fireball radius and heat flux.

**CONSTRAINTS**     Applicable to fireballs out-of-doors.

---

**The Method**

The empirical correlations derived by Moorhouse and Pritchard (1982), and Roberts (1982) are used. Both the radiation intensity at a distance from the fireball centre and the duration of the fireball can be determined using a very simple scaling law. The maximum radius of the fireball is given by:

$$R_f = 2.665 \, M^{0.327}$$

where M is the flammable release mass in kilogrammes.

The fireball has a duration, $t_f$, given by:

$$t_f = 1.089 \, M^{0.327}$$

The rate of release of energy by combustion is then given by:

$$Q = \frac{\eta \, H_c \, M}{t_f}$$

where $\eta$ is the combustion efficiency, found to vary with the saturated vapour pressure of the stored material (in $MN/m^2$) as follows:

$$\eta = 0.27 \, p_s^{0.32}$$

The radiation flux, I, at a distance r from the centre of the fireball is given by:

$$I = \frac{Q \, T}{4 \, \pi \, r^2}$$

where T is the transmissivity, conservatively taken as 1.

The above assumes that the rate of energy release is constant over the duration of the fireball; this assumption may be non-conservative. The radiation will, in fact, reach a peak which is greater than the average flux; the effects on people are more dependent on radiation level than on duration of exposure so the maximum effects may be missed if this peak radiation level is not used.

| | |
|---|---|
| Output | This method gives the radiation intensity at specified distances from the centre of the fireball. It also gives the maximum radius and duration of the fireball. |
| Input | Mass of release and heat of combustion. |
| Accuracy | These methods are based upon empirical correlations which may be updated from time to time in the light of new evidence. The methods are considered to be adequate for an initial hazard assessment using current state of the art techniques though the model may underestimate the effects on people. |
| Application | These methods may be used to estimate the radiation levels at a distance from the fireball. |
| Resource Requirements | These calculations can be performed with a calculator. |
| Linkage with Other Models | The heat flux from the fireball is used to calculate fire damage. |

## 4.4.4      Flash Fires

**METHOD**      The cloud dimensions and concentrations obtained from the gas dispersion models are used to estimate the area affected by the fire.

**OUTPUT**      Extent of flash fire

**CONSTRAINTS**      As for the dispersion models used to obtain cloud dimensions and concentrations

---

**The Method**

It is generally assumed that a flash fire spreads throughout the part of the vapour cloud which is above LFL. The dispersion calculations presented previously can be used to establish UFL and LFL contours. However, there is little information on the effects of a flash fire on people.

It is conservatively assumed that, if the cloud ignites, all the people who are out of doors and between the LFL and UFL contours will be killed; of the people inside buildings, a fraction, $F_{ki}$, will be killed. In an initial hazard assessment and in the absence of other information, $F_{ki}$ is usually taken to be zero.

For a detailed analysis the analyst should perform the calculations with ignition at different times after release, with different meteorological conditions, and with different wind directions. All of these factors have a considerable effect on the number of people within the boundaries of the cloud at the time of ignition.

---

**Outputs**

This method gives the extent of a flash fire and provides an estimate of the expected fatalities which is usually conservative.

**Inputs**

(i)  Cloud concentration profile from previous dispersion calculations.
(ii)  Flammability limits.
(iii) Population data for the area

**Assumptions**

The method assumes that significant over-pressures do not occur and includes rather crude assumptions about the protection provided by houses, and by being outside the flame.

**Accuracy**

Not known but considered to be adequate for the purpose of an initial hazard analysis.

**Application**

This method is applicable to a flash fire.

## 4.4.5     Fire Damage

**METHOD**     Fire damage estimates are based upon correlations with recorded incident radiation flux and damage levels

**OUTPUT**     Indication of damage as a function of incident radiation

**CONSTRAINTS**     Since damage estimates are based upon empirical evidence the damage response characteristics should be updated if new evidence comes to light.

---

**The Method**     There are various tables which give criteria for damage to people and property from fire. Sometimes they are expressed in terms of radiation intensity, and sometimes as a power dosage. The effect on people is expressed in terms of the probability of death and different degrees of injury for different levels of radiation. The effect on buildings, natural surroundings and equipment is measured in terms of the probability of ignition; this is particularly important for wooden structures. In Table 4.5 below, the radiative or incident flux is related to the levels of damage; this table is based on observations of large fires.

| Incident Flux ($kW/m^2$) | Type of Damage Caused | |
| --- | --- | --- |
| | **Damage to Equipment** | **Damage to People** |
| 37.5 | Damage to process equipment | 100% lethality in 1 min. 1 % lethality in 10s |
| 25.0 | Minimum energy to ignite wood at indefinitely long exposure without a flame | 100% lethality in 1 min Significant injury in 10s |
| 12.5 | Minimum energy to ignite wood with a flame; melts plastic tubing | 1% lethality in 1 min 1st degree burns in 10s |
| 4.0 | | Causes pain if duration is longer than 20s but blistering is unlikely. |
| 1.6 | | Causes no discomfort for long exposure. |

TABLE 4.5: *Damage Caused at Different Incident Levels of Thermal Radiation.*

---

| | |
|---|---|
| **Outputs** | The method provides estimates of fire damage, fatalities and injuries. |
| **Inputs** | Estimates of thermal flux at selected receptor points using appropriate fire models described in previous sections. |
| **Assumptions** | At lower levels of radiation, where time is required to cause serious injury to people, there is often the possibility to escape or take shelter. |
| **Accuracy** | The accuracy is considered adequate for initial hazard assessments. |
| **Application** | The correlations of thermally induced damage or injury may be applied to hazard assessment. |

## 4.5    EXPLOSIONS

An explosion is a sudden release of energy that causes damaging pressure waves in the atmosphere. The explosions associated with a chemical plant can take several forms:

a)   Explosive deflagration (or slow burning) of a dispersed flammable vapor. This is known as an Unconfined Vapor Cloud Explosion (UVCE).

b)   Detonation (or shock-wave burning) of an unconfined vapor cloud. This has never been definitely observed, but it is believed to be possible.

c)   Explosion of a flammable mixture in a confined space such as a building.

d)   Explosion of a pressure vessel due to a runaway reaction or other process abnormalities.

e)   Explosion of an unstable solid or liquid.

f)   Bursting of a pressurised container, with no chemical reaction involved.

The first five forms of explosion involve the release of chemical energy; the last involves the release of physical energy only. The effects of physical explosions tend to be local, but chemical explosions can have widespread effects. Because of this, most of the research on explosions and damage effects has concentrated on chemical explosions.

For chemical explosions a correlation has been devised which relates the distances to four different damage levels to the mass of fuel involved in the explosion. This should be used to obtain the medium- and far-field effect of all forms of chemical explosion on a chemical plant.

The near-field effects of an explosion should also be calculated, in order to determine whether the explosion will cause any "knock-on" effects in other equipment. This can be done using the correlation method mentioned above, but it can also be useful to calculate damage from projectiles. A method of relating effect distances and damage potential to the amount of energy released in the explosion. This method can be used for all pressure vessel explosions, both chemical and physical.

It should be noted that an explosion is only one of two possible results of the ignition of a flammable gas cloud. The other possible result is a flash fire, with radiation effects but no blast effects (Section 4.4.4). The probability split between the two events is a matter for the analyst's judgement; typically, 15% of delayed ignitions are assumed to result in explosions, and 85% in flash fires.

## 4.5.1    Explosion Correlation

**METHOD**        Correlation of damage produced with energy of explosion.

**OUTPUT**        Distances to various levels of damage caused by a vapour cloud detonation.

**CONSTRAINTS**   Should not be extrapolated for very large or very small clouds.

---

**The Method**

There are two methods of estimating the effects of explosions. The first is to estimate the damage levels directly and the second to find the overpressure and other parameters, and estimate damage from them. The former method is more direct and simple; it is therefore recommended here.

The correlation method given here is taken from TNO(1979). This method predicts damage radii R(s), given by:

$$R(s) = C(s) \left[ NE_e \right]^{\frac{1}{3}}$$

where C(s) is an experimentally derived constant defining the level of damage based on studies of the Flixborough (1974) and Beek (1975) vapor cloud explosions. The relationship between C(s) and the level of damage is given in Table 4.6 below.

$E_e$ is the total energy of the explosion, obtained by multiplying the heat of combustion by the mass of vapour within the flammable limits.

N is the yield factor, i.e. the proportion of $E_e$ which is available for pressure wave propagation. N is given by:

$$N = N_c \cdot N_m$$

where $N_c$ is the proportion of yield loss due to the continuous development of fuel concentration. Typically, $N_c$ is taken as 30%.

$N_m$ is the mechanical yield of combustion which is usually taken as 33% for constant volume or isochoric combustion (i.e. confined explosion), and as 18% for constant pressure or isobaric combustion (i.e. unconfined explosion). Most explosions are confined in some way so it is best to take $N_m$ as 33%, giving a total yield of 10%.

| C(s) | Limit Value $(\text{mJ}^{-1/3})$ | Characteristic Damage | |
|------|------|------|------|
| | | To Equipment | To People |
| C(1) | 0.03 | Heavy damage to buildings and to process equipment | 1% death from lung damage >50% eardrum rupture >50% serious wounds from flying objects |
| C(2) | 0.06 | Repairable damage to buildings and damage to the facades of dwellings | 1% eardrum rupture 1% serious wounds from flying objects |
| C(3) | 0.15 | Glass damage | Slight injury from flying glass |
| C(4) | 0.4 | Glass damage to about 10% of panes | |

TABLE 4.6: *Limit-values for Various Characteristic Types of Explosion Damage*

| | |
|---|---|
| Outputs | The method gives the distances to various levels of damage. |
| Inputs | In finding damage levels directly, it is assumed that the total amount of combustible material in the explosive part of the cloud and the heat release per unit mass of the material are known. The product of these gives the total energy $E_e$ of the explosion. |
| Assumptions | If the total energy available for explosion, $E_e$, is greater than $5 \times 10^{12}$ joules, there is almost no information on damage effects. For lower values $E_e$ there is sufficient data to make an estimate of damage. |
| Accuracy | It is generally considered that the correlation methods tend to give conservative results when applied to vapour cloud explosions. |
| Application | This method can be used to provide an estimate of effect distances for a range of explosion severities for flammable clouds of hydrocarbons containing up to $5 \times 10^{12}$ joules of energy, or approximately 100 tonnes of material. |

## 4.5.2 Damage from Projectiles

**METHOD**      Use of ballistics equations taking into account aerodynamic drag.  Also use of penetration equations.

**OUTPUT**      Range of projectile and assessment of potential to cause damage. .

**CONSTRAINTS**      This  model is appropriate for explosion of reactors and other large  process pressure vessels.

---

**The Method**

For  explosions from a closed vessel or in an area containing obstacles, fragments of the vessel or objects blasted by shock may become projectiles. It is possible to estimate the behaviour of these projectiles, considering energy, velocity, range and penetration of surrounding walls and equipment.

The  initial energy,  E,  of a projectile is considered to be a fraction of the explosion  energy; this fraction can be between 0.2 - 0.6.  The initial velocity $u_i$ can then be found from:

$$ E = \frac{1}{2} m u_i^2 $$

where m is the mass of the projectile.

The simple parabolic expression for the range of a projectile has been found to be inadequate, since air drag must be taken into account even for small masses. Various  empirical  equations are available for different types of projectile. Clancey (1972) gives for the range x:

$$ x = \frac{w^{\frac{1}{3}}}{k\,a} \ln\left[ \frac{u_i}{u} \right] $$

where   w  is the mass of the projectile
     k  is a constant which depends on whether the velocity of the projectile is subsonic (k = 0.0014) or supersonic (k = 0.002).
     a  is the drag coefficient, which is a function of fragment shape but is usually between 1.5 and 2.
     u  is the velocity on impact.

| Material | K | $n_1$ | $n_2$ |
|---|---|---|---|
| Concrete (crushing strength 35 MN/m$^2$) | $18. \times 10^{-6}$ | 0.4 | 1.5 |
| Brickwork | $23. \times 10^{-6}$ | 0.4 | 1.5 |
| Mild Steel | $6. \times 10^{-6}$ | 0.33 | 1.0 |

TABLE 4.7: *Parameters in Equation for Projectile Penetration Distance*

Equations for penetration are again dependent on size and shape of projectiles. For small masses Cox and Saville (1975) give penetration distance t as:

$$t = Km^{n_1} V^{n_2}$$

where m is the mass of the projectile.
V is the velocity on impact.
k, $n_1$ and $n_2$ are dependent on the nature of the material penetrated as shown in Table 4.7 above.

| | |
|---|---|
| Outputs | a) Initial velocity of fragment <br> b) Velocity of projectile a distance from the explosion <br> c) Penetration distance for projectile |
| Inputs | a) Energy of explosion <br> b) Mass of projectile <br> c) Estimate of fraction of explosion energy imparted to projectile |
| Accuracy | Incident data are sparse but tend to support the validity of this model. |
| Application | Explosion of all types of pressure vessels where internal explosion is initiating cause. |
| Resource Requirements | These calculations can be perfomed with a calculator. |

## 4.6 EFFECTS OF TOXIC RELEASES

The effects of exposure to a toxic material can be divided into two categories according to the duration and concentration of exposure:

a) **Acute Toxic Effects**
   These arise from short-term exposure at high concentrations. Carbon monoxide poisoning is an example of an acute toxic effect.

b) **Chronic Toxic Effects**
   These arise from long-term exposure at low concentrations. Asbestosis and lead poisoning are examples of chronic toxic effects.

Hazard assessments are usually concerned with releases of large quantities of material, which disperse after an hour or so; therefore, hazard assessments consider acute effects only.

The effects of acute exposure include:

a) **Irritation**
   Irritation can be felt by the respiratory system, by the skin, and by the eyes. With some materials the irritation occurs at low concentrations and can be a warning to people to seek shelter.

b) **Narcosis**
   Some materials affect people's responses, in such a way as to make them slow to seek shelter or to warn others.

c) **Asphyxiation**
   Most gases can cause asphyxiation by displacing oxygen from the atmosphere; others, such as carbon monoxide, cause asphyxiation by displacing oxygen from the blood and thus preventing oxygen from reaching the tissues.

d) **Systemic damage**
   Some materials cause damage to the organs of the body. This damage may be temporary, or it may be permanent.

The severity of the effects depends on the concentration and duration of exposure, and on the toxic properties of the material. Unfortunately, the toxic properties of many materials are not well-established because the data about their effects on humans are very sparse. It is difficult to derive properties from real incidents because the concentrations and durations in these incidents are usually not known. Most toxicity data are based on experiments on animals, in which concentrations and durations can be controlled. However, the applicability of the results of such experiments to humans is questionable, given the differences in body weight and physiology. Therefore the analyst should regard toxicity data with caution.

Another problem in predicting toxic effects is the variation in vulnerability among the population. Because of this variation, expressions for toxic effects give the proportion of the population which would be expected to manifest the effect after a given exposure. The above problems make it difficult to summarise the toxicity of materials in order to compare them, since duration, concentration, and percentage-affected all have to be standardised.

One way of comparing the toxicity of materials is to use the LC(50) for a stated duration of exposure; this is the Lethal Concentration which would be expected to kill 50% of the exposed population over the exposure period. Another way of comparing or expressing toxicity is to specify a harmless level of exposure for a given period as the limiting exposure criterion; these levels are known as Threshold Limit Values (TLVs) and are specified for periods such as a normal working life and short-term periods for emergency exposure. Yet another criterion is the IDLH value for a 30 minute exposure, as described in NIOSH (1978); this is the dose which is Immediately Dangerous to Life or Health at that exposure duration.

All of these criteria can be useful when choosing the lowest concentration of concern in dispersion calculations. This manual uses LC(50) in order to illustrate how toxic effects can be analysed but the analyst might want to use a different concentration; the analyst's chosen criterion can simply be substituted for the LC(50).

Toxicity data for some commonly used chemicals are given in Appendix C.

## 4.6.1      Effects of a Vapour Cloud of Toxic Gas

**METHOD**

The cloud dimensions and concentration obtained from the appropriate dispersion model are combined with population data and a dose/response relationship to predict toxic effects.

**OUTPUT**

Proportion of population affected.

**CONSTRAINTS**

The validity of this approach has not been fully developed for all toxic materials due to the paucity of relevant experimental data.

---

**The Method**

The early effects during the rapid discharge of material are not usually included in the toxicity calculations because these regions should be confined within the plant boundaries for most cases. Toxic effects during these stages are assumed to have little significance for the total toxic impact. Usually, the only toxic effects which are calculated are for the subsequent cloud dispersion phase. Using the appropriate model it is possible to calculate the cloud concentration, dimensions and location at different times. These are the data needed to calculate the toxic effect at various distances from the release point.

### The Probit Function

The toxic effect for an exposure to a given concentration and duration are obtained by using a probit equation. Probit equations are a widely-used way of expressing the probability of a stimulus causing an effect among a population. A probit is a probability unit lying between 0 and 10, which is directly related to probability as shown in Table 4.8. The probit Y can be related to concentration and duration of exposure by an equation of the form:

$$ Y = A_t + B_t \log_e \left[ C^n t_e \right] $$

where $A_t$, $B_t$ and n are properties of the toxic material

     C is the concentration of exposure

     $t_e$ is the duration of exposure

The quantitiy $C^n t_e$ is known as the toxic load.

$A_t$, $B_t$ and n are chosen for each toxic material so that the value of Y is a Gaussian distributed random variable with a mean value of 5 and a variance of 1.

As mentioned above, the analyst should regard the probabilities obtained by this method with caution; for many materials the values for $A_t$, $B_t$ and n are based on very limited data, or on data obtained from animal experiments.

| % Fatalities | 0 | 1 | 2 | 3 | 4 | 5 | 6 | 7 | 8 | 9 |
|---|---|---|---|---|---|---|---|---|---|---|
| 0 | - | 2.67 | 2.95 | 3.12 | 3.25 | 3.36 | 3.45 | 3.52 | 3.59 | 3.66 |
| 10 | 3.72 | 3.77 | 3.82 | 3.87 | 3.92 | 3.96 | 4.01 | 4.05 | 4.08 | 4.12 |
| 20 | 4.16 | 4.19 | 4.23 | 4.26 | 4.29 | 4.33 | 4.26 | 4.39 | 4.42 | 4.45 |
| 30 | 4.48 | 4.50 | 4.53 | 4.56 | 4.59 | 4.61 | 4.64 | 4.67 | 4.69 | 4.72 |
| 40 | 4.75 | 4.77 | 4.80 | 4.82 | 4.85 | 4.87 | 4.90 | 4.92 | 4.95 | 4.97 |
| 50 | 5.00 | 5.03 | 5.05 | 5.08 | 5.10 | 5.13 | 5.15 | 5.18 | 5.20 | 5.23 |
| 60 | 5.25 | 5.28 | 5.31 | 5.33 | 5.36 | 5.39 | 5.41 | 5.44 | 5.47 | 5.50 |
| 70 | 5.52 | 5.55 | 5.58 | 5.61 | 5.64 | 5.67 | 5.71 | 5.74 | 5.77 | 5.81 |
| 80 | 5.84 | 5.88 | 5.92 | 5.95 | 5.99 | 6.04 | 6.08 | 6.13 | 6.18 | 6.23 |
| 90 | 6.28 | 6.34 | 6.41 | 6.48 | 6.55 | 6.64 | 6.75 | 6.88 | 7.05 | 7.33 |

| | 0.0 | 0.1 | 0.2 | 0.3 | 0.4 | 0.5 | 0.6 | 0.7 | 0.8 | 0.9 |
|---|---|---|---|---|---|---|---|---|---|---|
| 99 | 7.33 | 7.37 | 7.41 | 7.46 | 7.51 | 7.58 | 7.58 | 7.65 | 7.88 | 8.09 |

TABLE 4.8 : *Transformation of Percentage Fatalities to Probits for Toxicity Calculations (Finney, 1971)*

### *Evaluating the Probit Expression*

To evaluate the probit, the toxic load, $C^n t_e$, must be calculated at positions of interest. At a given location the concentration will vary over time as the cloud passes and dilutes. The total toxic load for the location is obtained by considering different time steps and the average concentration during those time steps. Then for m time steps the total toxic load is given by:

$$\text{Total Toxic Load} = \sum_{i=1}^{m} C_i^n t_{ei}$$

This total toxic load is then used in the probit equation.

This method of summing the toxic loads is compatible with the results of the cloud dispersion models since the dispersion models give the concentrations and dimensions of the cloud at different time steps; this makes it simple to calculate the toxic load at a point for any time step. However, a computer analysis would be needed in order to consider the complete duration of the release and dispersion, and all the points of interest. Some methods of simplifying the analysis are suggested below. The object of these methods is to evaluate the distance from the release at which the toxic effect would have a value of 50%. The methods make use of the LC(50) concentrations, and evaluate the distance from the release at which the toxic effect would have a value of 50%.

Normally, the exposure time should be set equal to the release duration. However, even in emergency situations, where considerable confusion may prevail, an exposure duration of more than 30 minutes would probably be unrealistic since potential victims would tend to take avoiding or mitigating actions within this time.

## a) For a continuous release:

For a continuous release the concentration at a point is the approximately the same throughout the dispersion. This makes is very easy to calculate the toxic load.

For 50% probability of fatality the probit Y has a value of 5. Inserting this value in the probit function and rearranging gives:

$$\exp\left[\frac{5.0 - A_t}{B_t}\right] = C^n t_e$$

A value of C is then obtained from which the radius to 50% fatality may be estimated using the models described in Section 4.3.

## b) For an instantaneous release:

The toxic effect of an instantaneous release is much more difficult to assess because as the cloud passes over the population, the concentration at a given point will vary. It is not possible to give a simple method. Instead, the analyst must examine the concentration profile at locations of interest (e.g. nearest house, school) and calculate the toxic load for the time taken by the cloud to pass the location. This will give the probability of death at the location.

Outputs      This method gives the probability of experiencing a lethal (or injurious) dose
             of the toxic material at a given distance from a source, taking dispersion
             conditions into account.

Input        (i)  Toxicity parameters for the material under consideration.
             (ii) Concentration/time profiles of release.

Assumptions  Many assumptions are required for this model and it should be noted that the
             resulting estimates are intended to yield an indication of the effect distances
             associated with specified toxic hazards.  The calculations of dispersion and
             toxic effect are interlinked;  for complex situations, access to suitable
             computing facilities would be required.

Accuracy     This method is considered to have an accuracy no better than a factor 2.

Application  This method may be used to estimate approximate effect distances in the event
             of a toxic gas or vapour release.  These calculations may be based upon Probit
             relationships, LC(50),  IDLH or other relevant dose criterion for the toxic
             material.

# Chapter 5.

## Summary of Consequences

The results from the consequence models are in the form of a list of effect distances. The effect distances for a release are the maximum distances from the release point at which the effects of the release would be felt; the effects for the release could include death, different degrees of injury, and structural damage. If these results show that the hazard posed by the plant is unacceptable, then this hazard must be reduced. Possible methods of achieving this reduction are described in Chapter 6.

## 5.1  RESULTS FOR DIFFERENT WEATHER CONDITIONS

For the releases which reach off-site the analyst should repeat the calculations with different weather conditions, since windspeed and stability have a great effect on cloud dispersion. Stable weather gives the greatest effect distances so the analyst should consider the most stable weather conditions that occur at the site, as well as the most common weather conditions.

## 5.2  ORDERING AND PRESENTING THE RESULTS

A hazard analysis of even a small plant will produce a large number of results, especially if different weather conditions and ignition times are considered. Table 5.1 has been provided to assist in organising the results of the hazard analysis calculations. The analyst should use one sheet for each failure case or cluster of failure cases. These tables provide a convenient way of collating and storing the results.

The results should then be plotted as circles on suitable maps. Two such maps will be required: one should show the site in detail so that on-site effects can be plotted; the other map should cover an area of a few kilometers around the site so that off-site effects can be plotted.

### 5.2.1  Results for On-Site Hazard

The effect distances on the site map are useful for identifying the possibilities for knock-on effects. Knock-on effects occur when large, vulnerable inventories, load-bearing structures, and vital control equipment lie within the effect distances for structural damage. Structural damage can be caused by blast effects or by fire; the analyst should find out whether the equipment at risk has been protected by the provision of a blast wall or fire-fighting equipment.

The plotted effect distances can also be used to examine the adequacy of escape distances and routes, and the siting of highly-populated work areas.

### 5.2.2  Results for Off-Site Hazard

The effect distances on the large-scale map are used to determine the hazard which the plant poses to neighbouring residential communities and industrial installations; this is usually the main reason for conducting a hazard assessment. The hazard assessment will be especially concerned with risks from explosions, flash fires and toxic materials, since these have the potential to kill or injure at great distances from the site.

This map can also be used by plant management and designers, and government authorities when designing emergency response procedures.

## TABLE 5.1:  *Results Form*

| PLANT NAME | |
|---|---|
| ANALYST | |

### FAILURE DATA

| Unit Name | Unit Co-ordinates | Component Type | Failure Mode | Failure Size |
|---|---|---|---|---|
| | | | | |

### STORAGE CONDITIONS

| Material | Inventory | Phase | Pressure | Temperature |
|---|---|---|---|---|
| | | | | |

### EVENT TREE USED    (Tick appropriate boxes)

| Flammable Gas | Toxic Gas | Flammable Liquid | Toxic Liquid |
|---|---|---|---|
| | | | |

### MODELS USED    (Tick appropriate boxes)

| OUTFLOW | | | TRANSITION | | | DISPERSION | | | FIRE | | | | EXPLOSION | | TOXIC |
|---|---|---|---|---|---|---|---|---|---|---|---|---|---|---|---|
| Liquid | Gas | Two-Phase | Pool | Jet | Expansion | Dense | Neutral | Buoyant | Pool | Jet | Fireball | Flash | Blast | Projectile | |
| | | | | | | | | | | | | | | | |

### EFFECT DISTANCES OBTAINED FROM CONSEQUENCE ANALYSIS

| Type of Impact | Flammable | | | | Explosion | | | Toxic | |
|---|---|---|---|---|---|---|---|---|---|
| | Jet Fire R(37.5) | Pool Fire R(37.5) | Fireball R(37.5) | Flash Fire R(37.5) | Blast R(1) | Blast R(2) | Projectile | R(IDLH) | R(LC50) Duration: |
| Deaths | 50% | 50% | 50% | 50% | 100% | 50% | | | 50% |
| Structural Damage | Medium | Medium | Medium | Medium | Heavy | Medium | | None | None |
| Windspeed Stability | | | | | | | | | |
| | | | | | | | | | |
| | | | | | | | | | |
| | | | | | | | | | |
| | | | | | | | | | |

# Chapter 6.

## Hazard Reduction

By following the procedures in the previous chapters, the analyst will have identified which off-site areas are at risk from on-site failures, and also which failures could potentially lead to knock-on effects in the plant. The items of plant equipment responsible for the most severe of these consequences will have also been identified.

In this chapter some corrective measures which the analyst might consider to reduce the consequences, frequencies and impacts of the failures are proposed. These suggestions are necessarily of a general nature and the analyst must decide which are suitable for the process and site under study. It is not possible to be comprehensive and the emphasis is placed on basic design modifications rather than secondary "add-on" remedial measures.

## 6.1 REDUCTION OF CONSEQUENCES

The evaluation methods outlined in the previous chapters identify pieces of plant that have potential of causing severe damage both on and off-site. A list of measures leading to reduced effect distances are proposed here. The analyst should repeat the effect calculations to estimate the benefits of the proposed measures.

In order to reduce consequences it is sometimes necessary to make profound changes to the plant design. For this reason, a hazard analysis is best carried out at the design stage of the plant, when design, layout and siting modifications can be made relatively easily. However, even at the operational stage, a hazard analysis will reveal opportunities to reduce the consequences for a release. Some consequence-reduction possibilities include: reduction of inventories, modification of process or storage conditions, elimination of hazardous material, improvement secondary containment.

### 6.1.1 Reduce Inventories

The primary object should be to reduce the inventory of hazardous material, so that the potential off-site consequences of a release are greatly reduced or even eliminated. Inventories can be reduced by:

-   reducing the inventory of hazardous materials in storage and in the process. Many instances can be cited where it has been possible to operate plants with considerably lower quantities of raw materials and intermediate products than originally designed;

-   changing the process to produce the hazardous material as a small quantity of intermediate material. This removes the necessity to store large quantities of the material;

-   change from batch to continuous reaction system, with lower inventories, better mixing, etc.;

-   use a low inventory / high efficiency process, eg. distillation and evaporation systems.

## 6.1.2  Modify Process or Storage Conditions

If it is not possible to reduce the inventory of hazardous material, it may be possible to change the process or storage conditions to reduce the potential consequences of an accidental release.

-   store and process toxic gases in a suitable solvent rather than in large volumes;

-   store and handle all materials in small, discrete quantities rather than in large volumes;

-   process hazardous reactive materials in a large volume of recycle carrier-material containing the catalyst in a continuous reactor and thus prevent runaway reactions;

-   process hazardous material as a gas rather than as a liquid in a flammable solvent;

-   store hazardous gas as a refrigerated liquid rather than under pressure;

-   reduce process temperatures and pressures through process modifications.

## 6.1.3  Eliminate Hazardous Material

Should the first two alternatives prove ineffective, it may be possible to use material or alternative process routes to eliminate the hazardous material.

## 6.1.4  Improve Shut-down and Secondary Containment

If a release does occur it is possible to reduce the amount of material escaping from containment or from the site.

Automatic shut-down will reduce the amount of material escaping from containment, and also reduce the release duration.

There are also methods for keeping released material within the plant boundaries. These include:

-   water curtains to restrict gas releases;

-   improved design of dikes (or bunds).

    Dikes limit the spread of liquid in the event of a release so they prevent the liquid spreading outside the boundaries of the plant, and also reduce the evaporation rate by reducing the area of the pool.  In order for a dike to be effective it should have sufficient volume to contain all of the liquid, and should be shaped so as to prevent the liquid surging over the sides.  Surging can be prevented by using a floor that slopes away from the vessel, and having high, vertical walls.  The area should be as small as possible to reduce the evaporation rate, but there are several restrictions on the area: the wall should be far enough away from the vessel so that a jet release will not over-shoot it; and the wall should not be so high that it hinders fire-fighting.  The evaporation rate can also be reduced by making the dike from insulating concrete.

## 6.2  REDUCTION OF RISK

If it is not possible to reduce the consequences sufficiently using the above methods, it may be possible to reduce the risk (or probability) that a release will occur.  The risk of a release can be reduced by using reliability studies or techniques such as HAZOP to inprove the operation and control of the plant.  Risks can also be reduced by improving the maintenance and inspection of the plant.

## 6.3  REDUCTION OF IMPACTS

As well as reducing the consequences and risks of releases, the analyst should also consider measures to reduce the impact of these releases. Some possible measures are given below, arranged in order of increasing involvement of employees and people off-site:

- provision of bunkering or blast walls;

- firewalls/fire-proofing of structures;

- provision of escape routes for employees;

- provision of safety and emergency training for employees;

- implementation of emergency procedures on and off-site;

- provision of public alert systems and education of public;

- planning and training for evacuation;

- provision of safety buffer-zones around the plant boundary.

If these measures would not reduce the hazard to an acceptable level, it might be necessary to site the plant in a different area, with different meteorological conditions and population distribution.

# Appendices

# APPENDIX A

## Notation used in Chapter 4

### Section 4.1.1:   Liquid Outflow

| | | |
|---|---|---|
| $A_r$ | Area of release | $m^2$ |
| $C_d$ | Discharge coefficient | - |
| $C_{pl}$ | Specific heat of liquid at constant pressure | $J\,kg^{-1}\,K^{-1}$ |
| $F_{vap}$ | Fraction of liquid flashed to vapour | - |
| $g$ | Gravitational acceleration | $m\,s^{-2}$ |
| $h$ | Height of liquid in tank above discharge point | $m$ |
| $H_{vapb}$ | Enthalpy of evaporation at atmospheric boiling point | $J\,kg^{-1}$ |
| $P_a$ | Ambient pressure | $Nm^{-2}$ |
| $P_1$ | Process or reservoir pressure | $Nm^{-2}$ |
| $Q$ | Liquid release rate | $kg\,s^{-1}$ |
| $T_b$ | Boiling point | $K$ |
| $T_1$ | Temperature of liquid | $K$ |
| $\rho_1$ | Density of liquid | $kg\,m^{-3}$ |

### Section 4.1.2 :   Gas Outflow

| | | |
|---|---|---|
| $A_r$ | Area of release | $m^2$ |
| $C_d$ | Discharge coefficient | - |
| $M$ | Molecular weight | |
| $P_1$ | Process or reservoir pressure | $Nm^{-2}$ |
| $P_a$ | Ambient pressure | $Nm^{-2}$ |
| $Q$ | Release rate | $kg\,s^{-1}$ |

## Section 4.1.2 :    Gas Outflow  (continued)

| | | |
|---|---|---|
| R | Universal gas constant | $J\ mol^{-1}\ K^{-1}$ |
| $T_1$ | Temperature of liquid | K |
| Y | Outflow coefficient | |
| $\gamma$ | Specific heats ratio | - |

## Section 4.1.3:    Two Phase Outflow

| | | |
|---|---|---|
| $A_r$ | Effective open area of hole | $m^2$ |
| $C_d$ | Discharge coefficient | - |
| $C_{pl}$ | Specific heat of liquid at constant pressure | $J\ kg^{-1}\ K^{-1}$ |
| $F_{vap}$ | Fraction of liquid flashed to vapour | - |
| $H_{vap}$ | Enthalpy of vaporization | $J\ kg^{-1}$ |
| $P_1$ | Process or reservoir pressure | $Nm^{-2}$ |
| $P_c$ | Critical pressure | $Nm^{-2}$ |
| Q | Release rate | $kg\ s^{-1}$ |
| $T_c$ | Boiling point | K |
| $T_1$ | Temperature of substrate | K |
| $\rho_g$ | Density of gas | $kg\ m^{-3}$ |
| $\rho_1$ | Density of liquid | $kg\ m^{-3}$ |
| $\rho_m$ | Mean density of two phase mixture | $kg\ m^{-3}$ |

# Section 4.2.1:     Liquid Spread and Evaporation

| | | |
|---|---|---|
| a | Atmospheric stability coefficient | - |
| $a_s$ | Surface thermal diffusivity | $m^2\,s^{-1}$ |
| g | Gravitational acceleration | $m^2\,s^{-1}$ |
| $H_{vap}$ | Enthalpy of evaporation | $J\,kg^-$ |
| $H_0$ | Height above pool where there is no vapor | m |
| m | Mass | kg |
| $m_g$ | Evaporation rate from ground | $kg\,s^{-1}$ |
| $m_w$ | Mass transfer from wind | $kg\,s^{-1}$ |
| M | Molecular weight | |
| n | Atmospheric stability cofficient | - |
| $p_s$ | Surface pressure | $N\,m^{-2}$ |
| r | Radius of pool | m |
| $r_d$ | Maximum Pool Radius | m |
| $r_w$ | Radius at which evaporation rate equals liquid release rate | m |
| R | Universal gas constant | $J\,mol^{-1}K^{-1}$ |
| t | Time | s |
| $t_w$ | Time for Pool to Reach Radius $r_w$ | s |
| $T_a$ | Ambient temperature | K |
| $T_b$ | Saturation temperature of liquid | K |
| u | Wind speed | $m\,s^{-1}$ |
| W | Mass of material released | kg |
| $\beta$ | Spread rate parameter | |
| $\lambda_s$ | Surface coefficient of heat conduction | $W\,m^{-1}\,K^{-1}$ |
| $\rho_l$ | Liquid density | $kg\,m^{-3}$ |

## Section 4.2.2:     Jet Dispersion

| | | |
|---|---|---|
| $b_1$ | Shape parameter | - |
| $b_2$ | Shape parameter | - |
| $D_{eq}$ | Equivalent diameter | m |
| $D_0$ | Diameter of orifice | m |
| $c_m$ | Vapour concentratin on axis of the jet | $kg\ m^{-3}$ |
| $c_{x,y}$ | Concentration at point x,y | $kg\ m^{-3}$ |
| $c_w$ | Centre line concentration | $kg\ m^{-3}$ |
| $m_0$ | Mass flow rate of release | $kgs^{-1}$ |
| $u_m$ | Velocity on the axis at a distance x from the orifice | $ms^{-1}$ |
| $u_0$ | Real outflow velocity | $ms^{-1}$ |
| $\rho_{g,a}$ | Density of gas at ambient conditions relative to air at same conditions | $kg\ m^{-3}$ |
| $\rho_{g0,a}$ | Density of gas at out flow conditions,relative to air at ambient conditions | $kg\ m^{-3}$ |

## Section 4.2.3:     Adiabatic Expansion

| | | |
|---|---|---|
| $c_c$ | Core concentration | $kgm^{-3}$ |
| $c_{ce}$ | Final core concentration | $kgm^{-3}$ |
| $C_{p1}$ | Specific heat of liquid at constant pressure | $J\ kg^{-1}\ K^{-1}$ |
| $C_{pa}$ | Specific heat ofair at constant pressure | $J\ kg^{-1}\ K^{-1}$ |
| $C_v$ | Heat capacity at constant volume | $J\ kg^{-1}\ K^{-1}$ |
| $E$ | Energy of expansion | J |
| $F_{vap}$ | Fraction of liquid flashed to vapour | - |
| $F_{vap3}$ | Final fraction of liquid flashed to vapour | - |

# Section 4.2.3: Adiabatic Expansion (continued)

| | | |
|---|---|---|
| $H_{L(1)}$ | Initial enthalpy of liquid | J |
| $H_{L(2)}$ | Final enthalpy of liquid | J |
| $H_{vap}$ | Enthalpy of evaporation | $J\,kg^{-1}$ |
| $H_{vap(2)}$ | Final enthalpy of evaporation | $J\,kg^{-1}$ |
| $j_{ce}$ | Final core concentration | $kg\,m^{-3}$ |
| $K_d$ | Diffusion coefficient | - |
| $M$ | Mass of gas released | kg |
| $M_{air}$ | Mass of air in cloud | kg |
| $P_1$ | Initial pressure | $N\,m^{-2}$ |
| $P_a$ | Atmospheric pressure | $N\,m^{-2}$ |
| $r_c$ | Core radius | m |
| $r_{ce}$ | Final core radius | m |
| $r_{pe}$ | Peripheral core radius | m |
| $S_{L(1)}$ | Entropy of liquid during initial expansion stage | $J\,kg^{-1}K^{-1}$ |
| $S_{L(2)}$ | Entropy of liquid during final expansion stage | $J\,kg^{-1}\,K^{-1}$ |
| $S_{v(2)}$ | Entropy of vapour during final expansion stage | $J\,kg^{-1}\,K^{-1}$ |
| $t$ | Time | s |
| $T_1$ | Initial temperature | K |
| $T_2$ | Intermediate temperature | K |
| $T_3$ | Final temperature | K |
| $T_a$ | Temperature of air | K |
| $T_b$ | Saturation temperature of liquid | K |
| $U_1$ | Initial internal energy of gas | J |
| $U_2$ | Final internal energy of gas | J |
| $U_{L(1)}$ | Initial internal energy of liquid | J |
| $U_{L(2)}$ | Final internal energy of vapour | J |

## Section 4.2.3: Adiabatic Expansion (continued)

| | | |
|---|---|---|
| $V_1$ | Initial volume | $m^3$ |
| $V_2$ | Final volume | $m^3$ |
| $V_{cloud}$ | Cloud volume | $m^3$ |
| $V_{go}$ | Volume of gas at standard temperature and pressure | $m^3$ |
| $\rho_a$ | Density of air | $kg\ m^{-3}$ |
| $\rho_g$ | Density of gas | $kg\ m^{-3}$ |
| $\rho_l$ | Density of liquid | $kg\ m^{-3}$ |

## Section 4.3.1: Dense Cloud Dispersion

| | | |
|---|---|---|
| $C_f$ | Friction factor | - |
| $C_p$ | Specific heat at constant pressure | $J\ kg^{-1}\ K^{-1}$ |
| $g$ | Gravitational acceleration | $m\ s^{-2}$ |
| $H$ | Cloud height | $m$ |
| $H_0$ | Initial cloud height | $m$ |
| $h_n$ | Natural convection coefficient | - |
| $H_n$ | New cloud height | $m$ |
| $H_s$ | New cloud height | $m$ |
| $k$ | Lateral spreading coefficient | - |
| $L_0$ | Initial cloud length | $m$ |
| $L_n$ | New length of cloud | $m$ |
| $L_s$ | New length of cloud | $m$ |
| $q_f$ | Local heat flux | $J\ m^{-2}\ s^{-1}$ |
| $q_n$ | Natural heat flux | $kg\ s^{-1}$ |
| $Q_e$ | Entrainment rate | $kg\ s^{-1}$ |

## Section 4.3.1: Dense Cloud Dispersion (continued)

| | | |
|---|---|---|
| $R_0$ | Initial radius of cloud | m |
| Ri | Richardson number | - |
| $R_n$ | New radius of cloud | m |
| $R_s$ | New radius of cloud | m |
| $T_c$ | Temperature of the cloud | K |
| Tg | Temperature of the ground | K |
| $U_e$ | Entrainment velocity | m s$^{-1}$ |
| $u_l$ | Longitudinal turbulence velocity | m s$^{-1}$ |
| u' | Vector sum of wind and slumping velocities | m s$^{-1}$ |
| u* | Friction velcity | m s$^{-1}$ |
| $\alpha$ | Entrainment coefficient | - |
| $\gamma$ | Dense cloud edge mixing coefficient | - |
| $\rho_c$ | Density of cloud | kg m$^{-3}$ |
| $\rho_{c,a}$ | Density of cloud at ammbient conditions relative to air at same conditions | kg m$^{-3}$ |
| $\sigma_y$ | Dispersion parameter in the crosswind direction | m |

## Section 4.3.2: Dispersion of a Neutral-Density Cloud

| | | |
|---|---|---|
| a,b,c,d | Terms in expressions for dispersion parameters | - |
| c | Concentration of cloud | kg m-3 |
| $c_m$ | Centre line concentration of cloud | kg m-3 |
| Q | Mass instantaneously released at ground level | kg |
| Q* | Mass continuously released at ground level | kg |
| t | Time | s |
| u | Cloud speed | m s$^{-1}$ |
| $w_m$ | Cross wind width at matching point | m |

## Section 4.3.2:   Dispersion of a Neutral-Density Cloud (continued)

| | | |
|---|---|---|
| $w_v$ | Cross wind width at distance xv from matching point | m |
| $x$ | Distance from source point in x direction | m |
| $x_v$ | Upstream distance of starting point of cloud | m |
| $y$ | Distance from source point in y direction | m |
| $z$ | Distance from source point in z direction | m |
| $\sigma_x$ | Down-wind dispersion parameter | m |
| $\sigma_y$ | Cross-wind dispersion parameter | m |
| $\sigma_z$ | Vertical wind dispersion parameter | m |
| $\sigma_{ym}$ | Crosswind diispersion parameter at matching point | m |

## Section 4.4.1:   Pool fires

| | | |
|---|---|---|
| $C_p$ | Specific heat at constant pressure | $J\,kg^{-1}\,K^{-1}$ |
| $H$ | Flame height | m |
| $H_c$ | Heat of combustion | $J\,kg^{-1}$ |
| $H_{vap}$ | Heat of vaporization | $J\,kg^{-1}$ |
| $I$ | Intensity of heat radiation | - |
| $m$ | Mass | kg |
| $Q$ | Total heat flux | $W\,m^{-2}$ |
| $r$ | Pool radius | m |
| $R$ | Distance from pool centre | m |
| $t$ | Time | s |
| $T$ | Transmissivity of air path | - |
| $T_a$ | Ambient temperature | K |
| $T_b$ | Boiling point | K |
| $\rho_a$ | Density of air | $kg\,m^{-3}$ |
| $\eta$ | Efficiency factor | - |

## Section 4.4.2:    Jet Fires

| $H_c$ | Heat of combustion | $J\ kg^{-1}$ |
|---|---|---|
| $I$ | Intensity of heat radiation | $W\ m^{-2}$ |
| $n_p$ | Number of point sources | - |
| $Q$ | Release rate | $kg\ s^{-1}$ |
| $Q_p$ | Radiated heat | $J\ kg^{-1}$ |
| $r$ | Receptor distance | $m$ |
| $X_g$ | Emissivity factor | - |
| $\eta$ | Efficiency factor | |

## Section 4.4.3:    Fireballs and BLEVEs

| $H_c$ | Heat of combustion | $J\ kg^{-1}$ |
|---|---|---|
| $I$ | Intensity of heat radiation | $W\ m^{-2}$ |
| $M$ | Flammable release mass | $kg$ |
| $P_s$ | Saturated vapour pressure | $MN\ m^{-2}$ |
| $Q$ | Release of energy by combustion | $J$ |
| $R_f$ | Maximum radius of fireball | $m$ |
| $t_f$ | Fireball duration | $s$ |
| $T$ | Transmissivity | - |
| $\eta$ | Combustion efficiency factor | - |

## Section 4.5.1: Explosion Correlation

| | | |
|---|---|---|
| $C(s)$ | Damage coefficient | - |
| $E_e$ | Total energy of explosion | J |
| $N$ | Yield factor | - |
| $N_c$ | Proportion of yield loss | - |
| $N_m$ | Mechanical yield of combustion | - |
| $R(s)$ | Damage radius | m |

## Section 4.5.2: Damage from Projectiles

| | | |
|---|---|---|
| $a$ | Drag coefficient | - |
| $E$ | Initial energy of projectile | J |
| $k$ | Velocity constant | - |
| $K$ | Material constant | - |
| $m$ | Mass of projectile | kg |
| $n_1$ | Material constant | - |
| $n_2$ | Material constant | - |
| $t$ | Penetration distance | m |
| $u$ | Impact velocity | $m\ s^{-1}$ |
| $u_i$ | Initial velocity | $m\ s^{-1}$ |
| $V$ | Impact velocity of small mass | $m\ s^{-1}$ |
| $w$ | Mass of projectile | kg |
| $x$ | Range of projectile | m |

## Section 4.6.1: Effects of a Vapor Cloud of Toxic Gas

| | | |
|---|---|---|
| $A_t$ | Toxic material property coefficient | - |
| $B_t$ | Toxic material property coefficient | - |
| C | Concentration of exposure | $kg\ m^{-3}$ |
| n | Toxic material property coefficient | - |
| P | Probability of death | - |
| R | Radius of cloud | m |
| $t_e$ | Time of exposure | s |
| x | Distance from source point | m |
| Y | Probit probability unit | - |

# APPENDIX B

## World Bank Guidelines for
## Identifying, Analysing and Controlling
## Major Hazard Installations in
## Developing Countries

### Office of Environment and Scientific Affairs
### Projects Policy Department

### PREAMBLE

The European Community has taken a lead in developing guidelines controlling major accident hazards of certain industrial activities. The Environmental Council of the Community met on June 24, 1982 and adopted such a directive which member states were required to comply with by January 8, 1984.

Impetus was given to the European Community to consider the need to control major hazards by, in particular, four serious industrial accidents; the Flixborough explosion in 1974 killed 28 workers, injured 89 people and caused widespread damage to housing in the vicinity of the plant; the disaster at Beek in Holland in 1975, an explosion and fire killed 14 people on site following the release of propylene at the refinery. The two other cases were at Seveso and Manfredonia in Italy in 1976, where highly toxic substances were released contaminating the surrounding districts, and raising implications regarding the health of people exposed to the toxic releases.

Recently the explosion of natural gas in Mexico City killing some 450 people and the toxic gas release at Bhopal in India killing more than 2,500 people has highlighted the urgent need for the World Bank to adopt similar guidelines to those developed by the EEC. These latter two incidents illustrate the even greater risks that must be controlled in installations producing hazardous sutbstances in developing countries.

# TABLE OF CONTENTS

# 1.0 INTRODUCTION

1.  These guidelines are based substantially on the EC directive on the major accident hazards of certain industrial activities and regulations promulgated under the United Kingdom Health and Safety at Work Act. Parts of this guidelines have been quoted directly from "A Guide to Control of Industrial Major Accident Hazards Regulations 1984", UK Health and Safety Booklet HS (R)-21 (published by HMSO, London).

2.  Industrial activities involving certain dangerous substances have the potential to give rise to serious injury or damage beyond the immediate vicinity of the work place. These activities have commonly come to be known as "Major Hazards". These guidelines are concerned with the protection of the health and safety of persons in the workplace and persons outside the plant boundary, as well as the protection of the environment. Furthermore, they apply generally to industrial processes, storage and transport of hazardous material, but do not apply to nuclear or to extraction or mining operations or to licensed hazardous waste disposal sites. According to the guidelines, persons in control of activities involving certain dangerous, explosive, flammable and toxic substances must demonstrate that major accident hazards have been recognised, and that measures have been taken to prevent accidents and to control and minimise the consequences of those that do occur.

3.  It is the object of these guidelines to provide a framework in which a developer can supply evidence and justification for the safe operation of the proposed industrial activity. It is not the objective of these guidelines to provide details of specific methods of analysis, safe operating procedures, etc., which are the contents of the World Bank Manual of Industrial Hazard Assessment Techniques.

4.  In summary, these guidelines provide the criteria for identifying acutely toxic, flammable, explosive and reactive hazards, as well as providing an indicative list of these hazardous chemicals. In addition threshold quantities are specified, which require the developer to undertake a major hazard assessment and to implement measures to control major hazards that are identified in such an assessment.

# 2.0 POTENTIAL INDUSTRIAL HAZARDS

5.  Although 'major hazard (or major accident)' is defined in the guidelines and includes the phrase 'a major emission, fire or explosion', the definition uses a number of phrases which need to be interpreted. An occurrence will be a major accident if it meets the following conditions.

    (a)  that it leads to a serious danger to people or the environment;

    (b)  that it results from uncontrolled events in the course of an 'industrial activity'; and

    (c)  that it involves one or more 'dangerous substances'.

6.    'Serious danger to persons' should be taken to mean death or serious injury including to health, or the threat of death or serious injury, whether caused immediately by the accident (e.g., collapse of a populated building caused by an explosion) or as a delayed effect (e.g., pulmonary oedema following some hours after exposure to a toxic gas), and affecting or potentially affecting people inside and outside the installation.  It is emphasized that the accidents, actual or potential, should be major ones distinguished from other serious accidents not only by the severity of the casualties but by the number of them, or by the physical extent of the damage.

7.    The reference to delayed effects is not intended to include the cumulative effects of frequent exposure to small amounts of the dangerous substance and, therefore, brief excursions slightly above the routine control limits for toxic substances should not be considered as major accidents.

8.    'Serious danger to the environment' should be taken to mean a significant, relatively long-lasting (but not necessarily irreversible) effect on plants or animals on land, in the air or in the water which has the potential to lead to a serious danger to man.  For example, serious pollution by a toxic substance of a water course used for drinking water could pose a threat to man.

9.    'Uncontrolled developments' should be taken to mean that the occurrences of concern are likely to develop quickly; to be outside the normally expected range of operating problems; to present only limited opportunity for preventive action; and to require any such action to be in the nature of an emergency response.  It also serves to indicate that the guidelines are concerned with acute rather than chronic events, i.e., unconvenanted or unusual rather than covenanted or regular releases of the dangerous substances.  Similarly, 'a major emission' refers to a relatively large, sudden and unconvenanted release of the dangerous substance from its normal containment.

10.    It is clearly possible to identify, using a pragmatic approach, the installations and activities that pose the main threat of a major accident.  It is also relatively easy to decide whether an event was a major accident after it has occurred.  It is much less easy to define 'major accident' for the purpose of making predictions, as the developer is required to do in his major hazard assessment as an essential step in demonstrating the adequacy of the measures taken to prevent such accidents.  The following examples outline events which may be taken, prima facie, as major accidents:

   (a) any major fire giving rise to thermal radiation at the site or plant boundary exceeding $5 \ kW/m^2$ for several seconds;

   (b) any release actual or potential, of a hazardous substance where the total quantity released is a significant proportion of the quantity which invokes the guidelines, e.g., releases of kilogram quantities of Group A toxic substances; ton quantities of other toxic substances; ton quantities of pressurised or refrigerated flammable gases; or tens of tons of flammable liquid;

   (c) any vapour or gas explosion which could give rise to blast overpressures at the site or plant boundary exceeding 0.5 bar; and/or

   (d) any explosion of a reactive or explosive substance which could cause damage to buildings or plant outside the immediate vicinity sufficient to render them or it

11. These estimates should include all the quantities of each substance present whether it is in pure form or part of a mixture. However, the substances should be in a form capable of giving rise to major accident hazards, e.g., no account should be taken of ammonia unless it is anhydrous or is in water solution containing more than 50% by weight, nor should account be taken of stored chlorinated potable water.

12. For the purpose of these guidelines a "developer" is defined as a manufacturer, distributor, transporter or end-user, who manufactures, processes, stores, or transports hazardous chemicals in threshold quantities greater than those identified in Appendix II and III of this document.

# 3.0 IDENTIFICATION SYSTEM FOR MAJOR HAZARDS

## 3.1 Introduction

13. The proposed system for identifying major hazards is based on the quantity (or inventory) of hazardous substance stored or processed at an industrial site or in transit. In this context the term "installation" is used to describe the general activity which may result in a major accident hazard. The term "installation" is defined further in Appendix I.

14. The threshold quantities specified in Section 3.2 relate to the total quantity of substance held on site. A developer may be involved in an activity in which the quantity of a hazardous substance varies over a period, due either to seasonal demand, or because the site is complex and includes a number of processes each of which has an inventory which varies from day to day or even hour to hour. In such cases the maufacturer should make an estimate of the quantity of each substance liable to be on site and a decision as to whether the requirements of Section 4, 5 and 6, apply should be based on the maximum anticipated quantity.

15. The estimation of the quantity of a substance at a site should include all the amounts which are likely to be on the site under the control of the same manufacturer. In the case of production and process activities, this will include quantities in manufacture, use or processing, and associated storage (i.e., the storage that is used in connection with the process). Account should also be taken of any quantities in pipelines on the site and in internal transport operations. These estimates should include all quantities of any dangerous substance whether the substance is in pure form or part of a mixture or present as a by-product. For example, if a chemical plant manufactures a hazardous substance, the estimate of the total quantity should take account of the quantities which are present in reaction mixtures and purification processes together with the quantities which are in storage. In certain processes there may be circumstances where a significant quantity of a dangerous substance can only be produced if abnormal conditions develop in the plant. If such an event can reasonably be predicted, then this should be taken into account in estimating the overall quantity of the substance on a site. An example of this type of situation was the production of a significant quantity of TCDD (dioxin) when conditions of excess temperature and pressure developed in a plant producing 2, 4, 5-trichlorophenol at Seveso.

16. It should be noted that when estimating the quantity of hazardous substance to assess whether the site becomes subject to these requirements, it is necessary to add the quantity of substance in process to the quantity in associated storage together with any amount of the same substance in any installation within 500 metres owned by the same manufacturer. If the nearby installation is more than 500 metres away then it is only necessary to add the quantitites if these are such that there could be, in foreseeable circumstances, an aggravation of the major accident hazard. It is realised that in some cases interpretation of this 500 metre requirement might be difficult but reasons for inclusion or exclusion should be clearly specified.

17. Although an industrial activity involves, or is liable to involve, hazardous substances, this does not of itself make the activity subject to these World Bank guidelines. The hazardous substance must be present under circumstances which could give rise to a major accident. Thus, it may be possible for a developer to argue that a major accident cannot, in fact, arise. It may be that the physical state of a substance or the way in which it is distributed round the site may avoid the possibility of some types of accidents. This may be particularly relevant to many of the toxic substances which are involatile liquids or solids. These may not have the potential to cause a major accident unless some special factor, such as energy contained in a pressurised system, is present, although spillage into water courses may still remain a problem.

## 3.2    Threshold Quantities for a Major Hazard Assessment

18. The criteria for hazardous substances and the quantities above which a major hazard assessment is required by the World Bank are presented in Appendix II. This Appendix is divided into four sections, namely:

    (A)  very toxic substances;
    (B)  other toxic substances;
    (C)  highly reactive substances and explosives; and
    (D)  flammable substances.

19  The criteria for the two groups of toxic substances, namely groups (A) and (B) of the Schedule are given in terms of the toxic effects on populations of specified experimental animals, though in two cases in addition to fulfilling these criteria the substances must have physical and chemical properties capable of entailing major accident hazards. This is taken to mean that the properties are such that the toxic substance could be easily distributed throughout the environment if containment is breached, for example, a gas or highly volatile liquid or a solid which might be ejected from a pressurised reactor. A substance has to satisfy only one of the three criteria for ingestion, percutaneous (i.e., through the skin) or inhalation toxicity in the tables to qualify. In the case of Class (A) substances a major hazard assessment is required irrespective of the quantity involved.

20. In the case of Class (B) toxic substances quantities have been specified for some of the more common substances. However, for those unnamed substances which fall into the indicative criteria given in Appendix II, quantitites exceeding 1 ton would require a major hazard assessment.

21. A major hazard assessment also is required for any process using plant at a pressure greater than 50 bars when the product of the volume of the pressure system in cubic metres and the pressure in bars exceeds 10,000. Likewise for highly reactive chemicals (Class (C)) and flammable substances (Class (D)), threshold quantities are given in Appendix II which would require a major hazard assessment for the purpose of these World Bank Guidelines.

22. To assist in identifying major hazard chemicals an indicative list of the acutely toxic and reactive chemicals in Classes (A), (B) and (C) have been listed in Appendix III.

## 4.0 IMPLEMENTATION OF THE GUIDELINES

23. The guidelines require that proof of safe operation be available at any time. Developers must show that they have identified the major accident hazards arising from their acitivties and have taken adequate steps to prevent such major accidents in design, layout and siting, that they will provide adequate steps to prevent such major accidents during operations and will provide people on-site with the information, training and equipment to ensure their safety.

## 4.1 Requirements for a Major Hazard Assessment

24. For major hazard installations handling dangerous materials in excess of the quantities listed in Appendix II a major hazard assessment is required. This study must show that the activity will be carried on safely; it includes a description of the major accident hazards that could arise from a manufacturer's activities and the controls that are exercised to prevent them or to limit their consequences. 'Major accident' is defined in Section 2 and guidance is given on the definition in this section. The guidance that follows discusses some of the general issues that bear on the major hazard assessment.

### 4.1.1 Objectives

25. The objectives of the major hazard assessment are:

   a) to identify the nature and scale of the use of dangerous substances at the installation;

   b) to give an account of the arrangements for safe operation of the installation, for control of serious deviations that could lead to a major accident and for emergency procedures at the site;

   c) to identify the type, relative likelihood, and broad consequences of major accidents that might occur; and

   d) to demonstrate that the developer has appreciated the major hazard potential of the company's activities and has considered whether the controls are adequate.

26. In addition, the work that the developer does in preparing his assessment should enable him to provide the competent authority responsible for making emergency plans outside the installation with an estimate of the scale and consequences of the realisation of the hazards, in accordance with the requirements of these guidelines (see Section 5).

### 4.1.2 The Content of the Major Hazard Assessment

27. A major hazard assessment is essentially an abstract of relevant information about the major hazard aspects of the activities from a much more extensive body of information. This body of information will include plant design specifications, operating documents, maintenance procedures, and information derived from the examination of the major hazard potential by means of techniques such as described in the World Bank Manual of Techniques for Assessing Industrial Hazards.

28. The information required in the major hazard assessment falls into two broad categories: firstly, factual information about the site, its activities and surroundings, and secondly, (the core of the assessment) estimates of the scale of potential major accidents which occur at the installation and the means to prevent these hazards being realised.

29. It is not possible to specify precisely what the second part of the major jazard assessment should contain because the complexity of the potential hazards will vary greatly from site to site. However, the World Bank Manual provides detailed guidance on the methodology for carrying out such an assessment. The essence of the major hazard assessment, and the reason behind the choice of that term, is that the onus lies on the developer to assess his own hazards, take measures to control them adequately, and then to present his conclusions to the World Bank.

30. The major hazard assessment should, therefore, contain sufficient information about the major accident potential of the developer's acitivtes to enable judgement to be made whether the significant hazards have been identified and are being properly managed. In some instances it may be necessary to ask for information in the assessment to be supplemented by further information, but ideally the aim should be to provide an analysis which stands on its own as a demonstration that major accident hazards are being adequately controlled.

31. The major hazard assessment should provide adequate justfication for its conclusions, either by setting out the sources of the evidence for a particular argument, or by recording the principal assumptions in sufficient detail to enable them to be challenged if it emerges that they are critical to the conclusions of the assessment. For example, a major hazard assessment may state that the integrity of pressure vessels has been assured by the strict application of appropriate design codes, operating duties, maintenance and inspection procedures, in support of an assumption that the sudden failure of pressure vessels has been dismissed as a possible cause of a major accident. A major hazard assessment may also perhaps indicate that the risk of an aircraft crashing on the installation is insignificant in comparison with other causes of a major accident, because the site is well separated from the nearest airport and air traffic lanes. Clearly, the amount of evidence required on each aspect of the major hazard assessment will vary according to the importance of that aspect and in particular the consequence of the particular accident being considered.

### 4.1.3    Information to be Included in a Major Hazard Assessment

32.    The report shall contain the following:

### a)  *Information Relating to Susbstances listed in Appendix II and III:*

i)   The name of the substance as given in Appendix III or for substances included in Appendix II under a general designation, the name corresponding to the chemical formula of the substance.

ii)  A general description of the analytical methods available to the developer for determining the presence of the substance, or references to such methods in the sciencitific literature.

iii) A brief description of the hazards from the substance.

iv)  In cases where the substance may be isolated from process vessels, its percentage concentration, and the main impurities and their percentages.

v)   Where there is a potential for runaway reactions a full description is required such as given in the example in Appendix V, and the consequences of runaway reaction determined as part of the major hazard assessment.  This may involve computer modelling, as well as data from bench scale testing e.g. as detailed in References (8) and (10).

### b)  *Information Relating to the Installations:*

i)   A map of the site and its surrounding area to a scale large enough to show any features that may be significant in the assessment of the hazard or risk associated with the site.

ii)  A scale plant of the site showing the locations and quantities of all significant inventories of the hazardous substances.

iii) A description of the processes or storage involving the hazardous substance and an indication of the conditions under which it is normally held.

iv)  The maximum number of persons likely to be present on site.

v)   Information about the nature of the land use and the size and disribution of the population in the vicinity of the activity to which the major hazard assessment relates.

## c) Information Relating to the Management System for Controlling the Activity

i) The staffing arrangements for controlling the activity with the name(s) of the person(s), and if appropriate his (their) deputies or the competent body responsible for safety and authorised to set emergency procedures in motion and to inform outside authorities.

ii) The arrangements made to ensure that the means provided for the safe operation of the acitivity are properly designed, constructed, tested, operated and maintained.

iii) The arrangements for training of persons working on the site.

## d) Information Relating to the Potential Major Hazard Accidents:

i) A description of the potential sources of a major accident and the conditions or events which could be significant in bringing one about.

ii) A diagram of any plant or plants in which the activities are carried on sufficient to show the features which are significant as regards the potential for a major accident or its prevention or control.

iii) A description of the measures taken to prevent, control or minimise the consequences of any major accident.

iv) Methods of sizing and capacities of emergency relief systems, blow-down tanks, emergency scrubber systems, vent flares, etc., especially to handle two-phase flow conditions (i.e. the subject of a forthcoming American Institute of Chemical Engineers publication).

v) Information about the emergency procedures laid down for dealing with a major accident occurring at the site.

vi) Information about prevailing meteorological conditions in the vicinity of the site.

vii) An estimate of the number of people on site who may be particularly exposed to the hazards considered in the written report.

33. Further details on these items are given in Appendix IV, and the World Bank Manual of Industrial Hazard Assessment Techniques. A flow chart (Figure 1) summarises the procedures outlined above in Section 4.

**FIGURE B.1:** *Procedures for Major Hazard  Assessment And Control Of Major Hazard Installation*

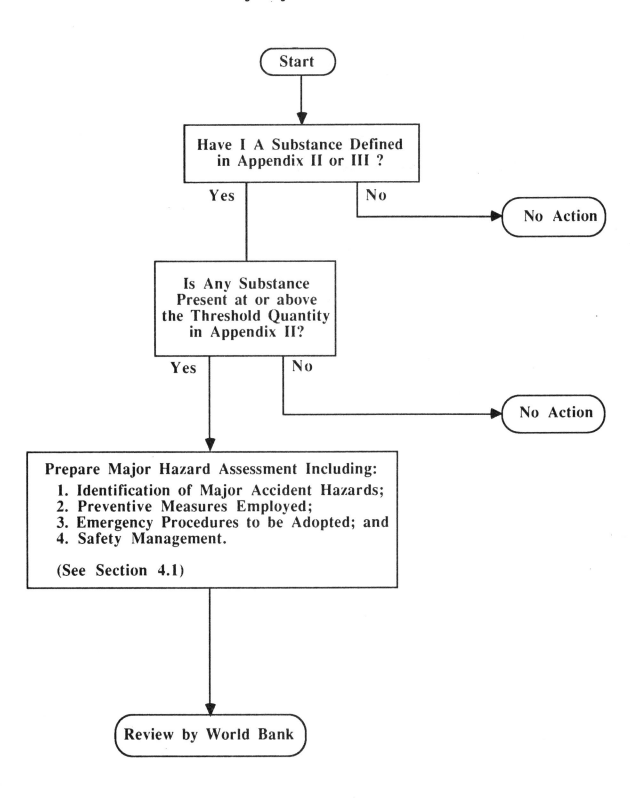

# 5.0 EMERGENCY PLANS

34. These guidelines require developers to prepare an adequate emergency plan for dealing with major accidents that may occur on their sites.

## 5.1 On-Site Emergency Plan

35. It is not the intention of these guidance notes to explain in detail how to prepare an on-site emergency plan. The detail and scope of the emergency plan will vary according to the complexity of the site and it is, therefore, not appropriate to prescribe here precisely what the plan should cover. The developer will need to consider the potential major accidents which are identified in the major hazard assessment (Section 4) to ensure that the plan takes account of them. Useful guidance in preparing the emergency plan may be found in the booklet 'Recommended Procedures for Handling Major Emergencies' published by the UK Chemical Industries Association, and the US EPA "Community Preparedness for Chemical Hazards, Part 3: A Guidance for Contingency Planning" (1985).

36. The developer should ensure that the on-site emergency plan is compatible with the off-site emergency plan which should be drawn up by the local authority. The on-site and off-site plans should be interlocked to ensure that they provide a comprehensive and effective response to emergencies.

37. The plan should include the name of the person responsible for safety on the site (usually the site or plant manager) and, if different, the name of the person who is authorised to set the plan in action.

38. The developer should keep the on-site emergency plan up-to-date, and to ensure that it takes account of any changes in operations on the site that might have a significant effect on the plan. The developer is also required to make sure that people on the site who are affected by the plan are informed of its relevant provisions. This should include not only those people who may have duties under the plan, but also those who may need to be evacuated from the site in an emergency, including contractors and visitors.

## 5.2 Off-Site Emergency Plan

39. The intention is that emergency plans should be drawn up or amended by the local authority after consultation with bodies who might be able to contribute information or advice. Such consultation is seen as an important aspect in the preparation of adequate emergency plans - this has been well demonstrated in the case of plans wich are in operation in many areas of the world. Obviously the developer must be consulted about the major accident hazards and the possible consequences, and any special emergency measures. The results of major hazard assessment discussed in Section 4 will provide useful data for drawing up these emergency plans.

40. A two-way flow of information is required betwen the developer and the local authority. Information from the developer is needed to enable the authority to draw up the off-site emergency plan; information from the authority should be available to the developer when he prepares the on-site emergency plan.

## 6.0 RESTRICTIONS ON DEVELOPMENT IN THE VICINITY OF MAJOR HAZARD INSTALLATIONS

41.  The extent of the safety buffer zone or restrictive development zone that may be required for a major hazard installation should be determined on a "case-by-case" basis. The importance of maintaining a restricted development, safety buffer zone is clearly shown by the experience in Mexico City and Bhopal, as well as at other hazard installations.

42.  The results of the major hazard assessment may indicate certain critical areas around the plant boundary where restrictions should be imposed on further development, taking into account local factors, as well as site storage, process and management factors etc. Planning authorities may restrict the land use in these safety buffer zones to warehousing, light industry or agricultural use, but exclude residential, shanty towns, hospitals, schools and commercial development. Some developers are purchasing safety buffer zones and are planting a dense tree cover as a safety screen, as well as to prevent squatting and shanty town developments.

# BIBLIOGRAPHY

1. European Community Directive, 1982. "On the Major Accident Hazards of Certain Industrial Acitivities". 82/501/EEC. Official Journal of the European Community, L230, June 1982.

2. "A Guide to the Control of Industrial Major Accident Hazards Regulations 1984." UK Health and Safety Executive HS (R) 21, HMSO (1985).

3. "UK Notification of Installations Handling Hazardous Substances Regulations, 1982." SI No. 1357 (and UK Health and Safety Executive Guide, Booklet HS (R) 16).

4. "A Guide to Hazard and Operability Studies". Chemical Industries Association, London, 1977.

5. "Recommended Procedures for Handling Major Emergencies", Chemical Industries Association, Alembic House, 93 Albert Embankment, London SE1.

6. "Codes of Practice for Chemicals with Major Hazards; Chlorine", Chemical Industries Association, 1975.

7. "Community Preparedness for Chemicals Hazards, Part 3; A Guide for Contingency Planning", US EPA (draft), August 1985, Document 6291H.

8. "Guidelines for a Reactive Chemicals Program", Dow Chemical Co., 1981.

9. "Techniques for Assessing Industrial Hazards: A Manual", World Bank, 1988.

10. "Daniel Stull, 'Fundamentals of Fire and Explosion'", AICHE Monograph Series No. 10, 73, 1977.

# APPENDIX I

## Definition of the Term "Installation"

1.  Installations for the production or processing of organic or inorganic chemicals using for this purpose, in particular:

    - alkylation

    - amination by ammonolysis

    - carbonylation

    - condensation

    - dehydrogenation

    - esterification

    - halogenation and manufacture of halogens

    - hydrogenation

    - hydrolysis

    - oxidation

    - polymerisation

    - sulphonation

    - desulphurisation, manufacture and transformation of sulphur-containing compounds

    - nitration and manufacture of nitrogen-containing compounds

    - manufacture of phosphorus-containing compounds

    - formulation of pesticides and of pharmaceutical products

2.      Installations for the processing or organic and inorganic chemical substances, using for this purpose, in particular:

- distillation

- estraction

- solvation

- mixing

- drying

3.      Installations for distillation, refining or other processing of petroleum or petroleum products.

4.      Installations for the total or partial disposal of solid or liquid substances by incineration or chemical decomposition.

5.      Installations for the production or processing of energy gases, for example, LPG, LNG, SNG.

6.      Installations for the dry distillation of coal or lignite.

7.      Installations for the production of metals or non-metals by the wet process or by means of electrical energy.

8.      Storage of dangerous materials identified in Appendices II and III.

9.      Transportation Distribution Systems:

- pipelines (quantities between block valves)

- shipping and terminal facilities (including in-land waterways)

- road

- rail

There may be an overlap between these guidelines and many national and international regulations, and guidelines concerning transfer of hazardous substances. When a national or international regulation applies to a particular guidelines should be used only as a check to ensure that all safety aspects have been identified and controlled.

# APPENDIX II

## List of Hazardous Substances Requiring
## a Major Assessment

### (A) "Very Acutely Toxic" Substances

The following indicative criteria are used to identify any "very toxic" substance requiring a major hazard assessment. These criteria are independent of the quantities of the substance stored, or processed, or that are formed by an unwanted by-product reaction.

Very toxic substances are defined as:

- substances which correspond to the first line of the table below;

- substances which correspond to the second line of the table below and wich, owing to their physical and chemical properties, are capable of entailing major-accident hazards similar to those caused by the substances mentioned in the first line:

|   | LD 50 (oral) (1) mg/kg body weightm | LD 50 (cutananeous) (2) g/kg body weight | LC 50 (inhalation) (3) mg/$l$ |
|---|---|---|---|
| 1 | LD 50 <5 | LD 50 <10 | LC 50 <0.1 |
| 2 | 5<LD 50<25 | 10<LD 50<50 | 0.1<LC 50<0.5 |

Note: (1) LD 50 oral in rats.
Note: (2) LD 50 cutaneous in rats or rabbits.
Note: (3) LC 50 by inhalation (four hours) in rats.

If an LC 50 value is available for a shorter exposure time "t" the LC 50 (4 hr) may be estimated as follows:

$$\text{LC 50 (4 hr)} = \frac{\text{LC 50 (t hr) x t}}{4}$$

## (B) Other Acutely Toxic Substances

(1)     The following quantities of toxic substances represent the threshold above which compliance with Section 4.1 is required.

| Named Substances | Quantity Tonnes |
|---|---|
| Phosgene | 2 |
| Chlorine | 10 |
| Hydrogen fluoride | 10 |
| Sulphur trioxide | 15 |
| Acrylonitrile | 20 |
| Hydrogen cyanide | 20 |
| Carbon disulpide | 20 |
| Sulphur dioxide | 20 |
| Bromine | 40 |
| Ammonia (anhydrous or as solution containing more than 50% by weight of ammonia) | 60 |

(2)     In addition to the above named substances, the following indicative criteria are used to identify other toxic substances which, owing to their physical and chemical properties, may cause a major accident and are stored or processed in quantities of greater than 1 tonne:

| LD 50 (oral) (1) (2) LC 50 (inhalation) (3) mg/kg body weight mg/l weight | | LD 50 (cutananeous) mg/kg body weight |
|---|---|---|
| 25 <LD 50 <200 | 50 <LD 50 <400 | 0.5 <LC 50<2 |

Note: (1) LD 50 oral in rats.
Note: (2) LD 50 cutaneous in rats and rabbits.
Note: (3) LC 50 by inhalation (four hours) in rats.

## (C) Highly Reactive Substances

(1)     The following quantities of "highly reactive" substances represent the threshold above which compliance with Section 4.1 is required.

| Named Substances | Quantity Tonnes |
| --- | --- |
| Hydrogen | 2 |
| Ethylene oxide | 5 |
| Propylene oxide | 5 |
| tert-Butyl peroxyacetate | 5 |
| tert-Butyl peroxyisobutyrate | 5 |
| tert-Butyl peroxymaleate | 5 |
| tert-Butyl peroxy isopropyl carbonate | 5 |
| Dibenzyl peroxydicarbonate | 5 |

| Named Substances | Quantity Tonnes |
| --- | --- |
| 2, 2-Bis(ter-butylperoxy) butane | 5 |
| 1,1-Bis(ter-butylperoxy) cyclohesane | 5 |
| Di-sec-butyl peroxydicarbonate | 5 |
| 2,2-Dihydroperoxydicarbonate | 5 |
| Di-n-propyl peroxydicarbonate | 5 |
| Methyl ethyl ketone peroxide | 5 |
| Sodium chlorate | 25 |
| Liquid oxygen | 200 |

| General Groups of Substances | Quantity Tonnes |
| --- | --- |
| Organic peroxides (not listed above) | 5 |
| Nitrocellulose compounds | 50 |
| Ammonium nitrates | 500 |

(2)     In addition to the above names substances, the following indicative criteria are used to identify potential explosive hazards, irrespective of materials stored or processed.

- Substances which may explode under the effect of flame or which are more sensitive to shocks or friction than dinitrobenzene.

## (D)  *Flammable Substances*

The following quantities of "flammable" substances represent the threshold, above which and compliance with Section 4.1 is required.

| Class of Flammable Substances | Quantity Tonnes |
|---|---|
| **1.** *Flammable Gases:* | |
| Gas or any mixture of gases which is flammable in air and is held as a gas. | 15 |
| **2.** *Liquefied Gases and Flammable Liquids in Process at Pressure and/or Temperature Above Ambient Levels:* | |
| A substance or any mixture of substances which is flammable in air and is normally held in the installation above its boiling point (measured at 1 bar absolute) as a liquid or as a mixture of liquid and gas at a pressure of more than 1.4 bar absolute.  (e.g.  LPG's.) | 25 being the total quantity of substances above the boiling points whether held straight or in mixtures. |
| **3.** *Refrigerated Liquefied Gas:* | |
| A liquefied gas or any mixture of liquefied gases, which is flammable in air, has a boiling point of less than 0°C (measured at 1 bar absolute) and is normally held in the installation under refrigeration or cooling at a pressure of 1.4 bar absolute or less (e.g.  LNG). | 50 being the total quantity of substances having boiling points below 0°C whether held singly or in mixtures. |
| **4.** *Highly Flammable Liquids:* | |
| A liquid or any mixture of liquids not included in items 1 to 3 above, which has a flash point of less than 21°C. | 10,000 |

**5.** *Flammable Liquids at  High Temperature and Pressure:*

Substances which have a flash point lower than 55°C and which remain liquid under pressure, where particular processing conditions, such as high pressure and temperature, may lead to a major accident hazard.

# APPENDIX III

## List of Acutely Toxic and Reactive Hazardous Substances

**Group A: Very Toxic Substances (See definition Appendix II)**

Aldicarb

4-Aminodiphenyl

Amiton

Anabasine

Arsenic pentoxide, Arsenic (V) acid and salts

Arsenic trioxide, Arsenious (III) acid and salts

Arsine (Arsenic hybride)

Azinphos-ethyl

Azinphos-methyl

Benzidine

Benzidine salts

Beryllium (powders, compounds)

is (2-chloroethyl) sulphide

Bis (chloromethyl) ether

Carbofuran

Carbophenothion

Chlorfenvinphos

4-(Chloroformyl) morpholine

Chloromethyl methyl ether

Cobalt (powders, compounds)

Crimidine

Cyanthoate

Cycloheximide

Demeton

Dialifos

00-Diethyl S-ethylsulphinylmethyl phosphorothioate

00-Diethyl S-ethylthiomethyl phosphorothioate

00-Diethyl S-isopropylthiomethyl phosphorodithioate

00-Diethyl S-propylthiomethyl phosphorodithioate

Dimefox

Dimethylcarbamoyl chloride

Dimethylnitrosamine

Dimethyl phosphoramidocyanidic acid

Diphacinone

Disulfoton

EPN

Ethion

Fensulfothion Fluenetil

Fluoroacetic acid

Fluoroacetic acid, salts

Fluoroacetic acid, esters

Fluoroacetic acid, amides

4-Fluorobutyric acid

4-Fluorobutyric acid, salts

4-Fluorobutyric acid, esters

4-Fluorobutyric acid, amides

4-Fluorocrotonic acid

4-Fluorocrotonic acid, salts

4-Fluorocrotonic acid, esters

4-Fluorocrotonic acid, amides

4-Fluoro-2-hydroxybutyric acid

4-Fluoro-2-hydroxybutyric acid, salts

4-Fluoro-2-hydroxybutyric acid, esters

4-Fluoro-2-hydroxybutyric acid, amides

Glycolonitrile (Hydroxyacetonitrile)

1, 2, 3, 7, 8, 9-Hexachlorodibenzo-p-dioxin

Hexamethylphosphoramide

Hydrogen selenide

Isobenzan

Isodrin

Juglone (5-Hydroxynaphthalene-1,4-dione)

4,4'-Methylenebis (2-chloroaniline)

Methyl isocyanate

Mevinphos

2-Naphthylamine

Nickel (powders, compounds)

Nickel tetracarbonyl

Oxydisulfoton

Oxygen difluoride

Paraoxon (Diethyl 4-nitrophenyl phosphate)

Parathion

Parathion-methyl

Pentaborane

Phorate

Phosacetim

Phosphamidon

Phosphine (Hydrogen phosphide)

Promurit (1-(3,4-Dichlorophenyl)-3-triazenethiocarboxamide

1,3-Propanesultone

1-Propen-2-chloro-1,3-diol diacetate

Pyrazoxon

Selenium hexafluoride

Sodium selenite

Stibine (Antimony hydride)

Sulfotep

Sulphur dichloride

Tellurium hexafluoride

TEPP

2, 3, 7, 8-Tetrachlorodibenzo-p-dioxin (TCDD)

Tetramethylenedisulphotetramine

Thionazin

Tirpate (2,4-Dimethyl-2,3-dithiolane-2-carboxaldehyde 0-methylcarbamoyloxime)

Trichloromethanesulphenyl chloride

1-Tri(cyclohexyl)stannyl-1H-1,2,4-triazole

Triethylenemelamine

Warfarin

## Group B: Other Toxic Substances (See definition Appendix II )

Acetone cyanohydrin (2-Cyanopropan-2-ol)

Acrolein (2-Propenal)

Acrylonitrile

Allyl alcohol (2-Propen-1-ol)

Allylamine

Ammonia

Bromine

Carbon disulphide

Chlorine

Ethylene dibromide (1,2-Dibromoethane)

Ethyleneimine

Formaldehyde (concentration > 90%)

Hydrogen chloride (liquefied gas)

Hydrogen cyanide

Hydrogen fluoride

Hydrogen sulphide

Methyl bromide (Brommethane)

Nitrogen oxides

Phosgene (Carbonyl chloride)

Propyleneimine

Sulphur dioxide

Tetraethyl lead

Tetramethyl lead

## Group C.1: Highly Reactive Substances and Explosives

Acetylene

Ammonium nitrate (where it is in a state which gives it properties capable of creating a major accident hazard)

2,2-Bis(tert-butylperoxy) butane (concentration > 70%)

1,1-Bis (tert-butylperoxy) cyclohexane (concentration > 80%)

tert-Butyl peroxyacetate (concentration > 70%)

tert-Butyl peroxyisobutyrate (concentration > 80%)

tert-Butyl peroxy isopropyl carbonate (concentration > 80%)

tert-Butyl peroxymaleate (concentration > 80%)

tert-Butyl peroxyphivalate (conentration > 77%)

Dibenzyl peroxydicarbonate (concentration > 90%)

Di-sec-butyl peroxydicarbonate (concentration > 80%)

Diethyl peroxydicarbonate (concentration > 30%)

2,2-Dihydroperoxypropane (concentration > 30%)

Di-isobutyryl peroxide (concentration > 50%)

Di-n-propyl peroxydicarbonate (concentration > 80%)

Ethylene oxide

Ethyl nitrate

3,3,6,6,9,9-Hexamethyl-1,2,4,5-tetroxacyclononane (concentration > 75%)

Hydrogen

Methyl ethyl ketone peroxide (concentration > 60%)

Methyl isobutyl ketone peroxide (concentration > 60%)

Peracetic acid (concentration 60%)

Propylene oxide

Sodium chlorate

## Group C.2: Explosive Substances

Barium azide

Bis (2,4,6-trinitrophenyl) amine

Chlorotrinitrobenzene

Cellulose nitrate (containing > 12.6% nitrogen)

Cyclotetramethylenetetranitramine

Cyclotrimethylenetrinitramine

Diazonidinitrophenol

Diethylene glycol dinitrate

Dinitrophenol, salts

Ethylene glycol dinitrate

1-Guanyl-4-nitrosaminoguanyl-1-tetrazene

2,2',4,4',6,6'-Hexanitrostilbene

Hydrazine nitrate

Lead azide

Lead styphnate (Lead 2,4,6-trinitroresorcinoxide)

Mercury fulminate

N-Methyl-N,2,4,6-tetranitroaniline

Nitroglycerine

Pentaerythritol tetranitrate

Picric acid (2,4,6-Trinitrophenol)

Sodium picramate

Styphnic acid (2,4,6,-Trinitroresorcinol)

1,3,5-Triamino-2,4,6-Trinitrobenzene

Trinitroaniline

2,4,6-Trinitroanisole

Trinitrobenzene

Trinitrobenzoic acid

Trinitrocresol

2,4,6-Trinitrophenetole

2,4,6-Trinitrotoluene

# APPENDIX IV

## Background Information Required
## for a Major Hazard Assessment
## (as discussed in Section 4.1)

Section 4.1.3 specifies the details required to be included in a major hazard assessment prepared in accordance with these guidelines.

The details required in Section 4.1.3 are discussed below as separate items but it is for the developer to determine the most appropriate or convenient method of presenting the required information. In particular, general issues (e.g., pressure vessel inspection arrangements) could be referenced and any variations or departures from the generally accepted practice will suffice.

The item-by-item guidance given below in relation to Section 4.1.3 is by example rather than by lists of topics to be covered. The latter approach suffers the twin disadvantages of not being truly comprehensive while at the same time giving the impression that each listed topic is of equal importance. In practice, the depth of information required on each topic will vary according to the circumstances of the individual installation.

Sections 4.1.3 (a) and 4.1.3 (b) require factual information about the dangerous substances and the installation handling them. Section 4.1.3 (c) relates to the management control of the activity. Section 4.1.3 (d) requires information about the sources and nature of potential major accidents and the measures taken to prevent and control them.

## IV.1 INFORMATION ON HAZARDOUS SUBSTANCES
### (Section 4.1.3 (a))

### (i)  Substance Name

The information required under this sub-heading is concerned with identifying the dangerous substances which qualifies the activity or storage for the requirement to make a major hazard assessment the working should be self-explanatory.

### (ii)  Monitoring Methods

Where standard analytical methods are used by the firm this item need only identify the method and any departures from it. Provision of gas detection equipment could be referred to here or left if appropriate to the discussion of preventive measures under Section 4.1.3 (d).

## (iii)  Hazards of the Substances

Information given under this sub-item should cover (a) the route of the harm to man (skin contact, inhalation or ingestion for toxic substances, and flame contact, thermal radiation or blast for flammable or explosive substances); (b) the dose-response relationship, where known, citing standard published references as appropriate (e.g., the Chemical Industries Association's table of the toxic effects of chlorine at various concentrations); and (c) the nature of the trauma, where this is not obvious (e.g., chloracne from exposure to dioxin).

In relation to the hazards from the substance to the environment, information is not required about the obvious effects of flame or blast from explosive or flammable substances. For toxic substances such information as is readily available should be given or be referred to, including the route of the harm (e.g., the pollution of water courses); the effect on flora and fauna which may be exposed; and an indication of the substance's persistence.

## (iv)  Composition of Process Streams

This item requires information about the composition of the substance so that the effects of diluents or impurities which have a significant effect on the hazard can be assessed. For example, where an organic peroxide is present in an acitivity, the name and percentage of any stabiliser should be given.

The presence of significant quantities of impurities could also affect the behaviour of a substance and these impurities and their percentages should also be identified. An example would be the alteration of the toxic properties of chloromethyl methyl ether by the presence of bis (chloromethyl) ether. It is not necessary to list the minor components of mixtures being processed where these have an insignificant effect on the potential hazard (e.g., a hydrocarbon mixture might be described as 70% butane, 25% propane, 5% higher hydrocarbons).

## IV.2  INFORMATION ON THE INSTALLATION
### (Section 4.1.3 (b))

## (i)  Location

This map is required to indicate where the installation is located, showing its position in relation to local geographic features, such as roads and towns. In general it will be sufficient to use the latest available map or maps on a scale which includes both the site and the surrounding features. Changes (such as a new motorway) which have occurred since the printing of the map and which are known to the firm should be shown; it is not intended that this should involve any extended research effort. For many sites a scale of 1 to 10,000 will be appropriate.

## (ii)    Plot Plan

The intention of this item is to identify clearly, both in location and quantity, the main parts which contribute to the total inventory of the dangerous substance. In addition, the plan should be annotated to indicate the lesser quantities which make up the stated total. For example, an estimate of the quantity of the substance which is present in pipework around a particular plant should be made.

It may be convenient to combine the response to this item with that to Section 4.1.3 (d), and in particular, Section 4.1.3 (d) item (ii).


## (iii)    Process/Flow Description

The intention of this requirement is that a sufficient description of the process be given to enable discussion on later items to be understood and placed in context. The amount of description required will depend on the complexity of the process. For example, a water treatment plant using chlorine will require only a brief account of the water dosing process, whereas a chemical plant producing qualifying quantities of a very toxic substance as an intermediate will require sufficient information to enable the critical aspects of the process chemistry to be understood. In this latter example, the information will be supplemented by the information given under Section 4.1.3 (d) and it will be for the developer to determine the most appropriate method of presentation.

The conditions under which the substance is normally held should be stated, including the physical state and pressure and/or temperature at the main stages of storage and process, e.g., butane is held in storage as a refrigerated liquid at $0^{o}C$; vaporised in a direct-fired evaporator and fed as a gas to a process vessel at 10 bar, $90^{o}C$.


## (iv)    Personnel on Site

The number given should be accompanied by sufficient explanation to show how it was derived. Account should be taken of the number of people who may be present at shift changeover; people who may be employed from the site but who may be present only for short periods (e.g., sales staff or delivery drivers); regular visitors to the site (e.g., contractors); and casual visitors. Exact numbers are not required.

## (v)    Local Land Use and Population Distribution

The information required is about the use of land or water surrounding the activity and the location of people who may be affected in the event of a major accident. This can be provided by annotations on a suitable map, indicating broad categories of land use (e.g., dwellings, other factories, schools, sports facilities, agricultural land, etc.). It is not necessary to give numerical estimates of population in the area covered by each category, though any unusually high densities such as shanty towns, blocks of flats should be marked as such.

The choice of the phrase 'in the vicinity of' rather than definite distances is intended to allow for flexible interpretation in relation to the potential hazard. For example, for a flammable liquid tank farm the 'vicinity' might be less extensive than for bulk chlorine storage. A degree of judgement is thus called for and a brief explanation should be given of the choice of vicinity illustrated on the map.

## IV.3    INFORMATION REQUIRED ON MANAGEMENT SYSTEMS  (Section 4.1.3 (c))

The aim of this item is to demonstrate that the developer has a proper management system and technical staff to control the major hazards aspects of his activities. The required information is general, insofar as it relates to the management control of the site as a whole, in contrast to the greater detail that may be needed under Section 4.1.3 (d) to demonstrate that there are adequate arrangements to prevent control particular hazardous events. The extent of the response to this item should be seen against its importance in providing a framework in which the rest of the safety case may be set and which will to some degree colour the credibility of the whole submission.

## (i)    Responsible Person and Staffing

A description of the management structure should be given which covers reporting relationship and the experience and qualifications of staff at the different levels. It is important to show how accountability for decisions which affect the potentially hazardous activity is assigned to staff who have the appropriate level of expertise and the relevant professional discipline. Reference should also be made to the developer's policy towards the appointment of competent deputies to cover key positions.

This section should also cover the arrangements which management have set up for identifying and dealing with safety issues arising from the potentially hazardous activity, with references as appropriate to the group, division and site safety policies, and the role of safety representatives and safety committees. This section should include an account of how management decisions about the potentially hazardous activity are made with due regard for their safety implications, and how these decisions are monitored.

## (ii)    Quality Control for Safety

The importance of the containment of dangerous substances is such as to require an adequate account of the management of the engineering system. This should include a discussion of the developer's approach to the design of important plant items and safety systems (e.g., use of standard or company codes); the arrangements for quality assurance; the inspection and testing procedures (who carries them out? how frequently? who decides on the necessary action in the event of deficiencies being discovered? who monitors these actions and how?); the provision of operating manuals and instructions and the procedures for revising them in the event of process changes; the procedures for ensuring that plant modifications are adequately designed, installed and tested; and the system for general maintenance.

The response to this item should not contain lengthy abstracts from written company procedures, but should aim to give a coherent justification for the system of engineering controls exercised over the potentially hazardous activity.

## (iii)    Training

Information should be given about the standards of relevant training, both on- and off-the-job, for people on site with a significant role in the control or operation of the major hazard activity, including the senior management and engineering staff involved. A brief account should also be given about how training needs are identified and met.

## IV.4    INFORMATION ON POTENTIAL MAJOR ACCIDENTS
(Section 4.1.3 (d))

The response to this item will provide the heart of the major hazard assessment. Though the information required is specified in some detail, it will be essential for developers to interpret this item with commonsense and flexibility. As noted in the general discussion of the term, "major hazard assessment", the developer must investigate those potential major accidents which may produce adverse consequences outside the plant boundary using hazard analysis techniques such as described in the World Bank Manual (ref. (9)). Sufficient evidence must be provided to support his conclusions.

## (i)    Identification of Potential Major Hazard Events

This item requires the manufacturer to identify the ways in which a major accident might occur. "Major accident" is defined in Section 2 and guidance on its interpretation is given in this section. Suitable techniques for identification are hazard analysis, hazard and operability studies, reactive chemicals studies, engineering flow chart review, and review of past accidents and near misses etc. The concept of the major hazard assessment allows the developer the opportunity to argue in the response to this item that his activities are not capable of giving rise to a major accident, provided that they are satisfied that such an argument can be fully and properly justified.

In some cases, a submission of this sort may be relatively easy to support. For example, a toxic hazardous substance may be stored in separate cylinders, and the quantity in each one may not be sufficient to give rise to a major accident. If the cylinders were disposed around the site so that the possibility of an incident affecting them simultaneously could be ruled out, then the argument that a major accident cannot arise on the site may be reasonably straightforward. In other cases the evidence necessary to justify such an argument may provide to be as extensive as a response to this item which accepts that major accidents may occur, and then goes on to describe how they are prevented; it will be for the manufacturer to decide initially which option better fulfills the objectives of the major hazard assessment. If it can be shown that the acitivity is not capable of giving rise to a major accident, then the remaining parts of Section 4.1.3 (d) need not be answered. It is in this section that the quantification of the magnitude of hazardous events by means of hazard analysis techniques are required.

In most cases cases it will be necessary to describe the potential sources of a major accident (the World Bank document "Techniquies for Assessing Industrial Hazards : A Manual" provides guidance in this respect). Storage and process vessels which contain significant quantities of the dangerous substance should be examined for the most probable ways in which their inventories may be released, and these should include consideration of, spontaneous failure (due to original defects or those arising in the course of operation); failure due to excursions from normal operating conditions (including such matters as operator error, loss of services, and failure of control devices); failure due to events elsewhere on site (e.g., fire, explosion); and failure due to external events (e.g. flooding, seismic activity). For a complex chemical or petrochemical works or a refinery, where the number of vessels and pipes for individual plants may be very large (but built to common standards), it may be appropriate to be more selective in examining the potential sources of a major accident by considering only the largest vessels in detail and then referring to smaller vessels or groups of vessels in general terms.

## (ii)     Process Flow Diagram

This item calls for judgment by the developer as to which sections of the plants containing the dangerous substance need to be illustrated in diagrammatic form. Such diagrams should show the process vessels, storage facilities and instrumentation on the plants concerned in sufficient detail to enable the discussion under Section 4.1.3 (d) (i) and (iii) to be readily understood. For vessels in the plants which have been identified in Section 4.1.3 (d) (i) as significant as regards the potential for a major accident, details should be shown on the diagram of their designed maximum working capacities, their design temperatures and pressures, and their normal operating conditions. It may be convenient to combine this with the response to Section 4.1.3 (b) (ii). For example, a large vertical cylindrical storage vessel containing the dangerous substance would require a diagram showing the vessel, its connections, instrumentation and external safety features such as water sprays and bunds; for complex plants diagrams of only the largest or most significant vessels are required.

## (iii)   Preventive and Control Systems

The third part of Section 4.1.3 (d) requires a response in two parts.  First, relating to the preventive and control measures which check any sequence of events which could foreseably result in a major accident, and secondly, relating to measures which may be taken after such a hazardous outcome to minimise its adverse consequences.  It is important to realise that it is better to prevent a release occurring rather than trying to control the consequences, which may, in part, be impracticable.

The first part of the response should concentrate on those preventive or control measures which are critical in counter-acting significant hazards, though many of these measures will also be appropriate to preventing or controlling lesser events.

The interpretation of certain key words in this part of Section 4.1.3 (d) may usefully be discussed in relation to examples; say, a large pressure storage sphere of LPG, representing the risk of fire and explosion, and for an example of toxic risk, a large pressure storage sphere of ammonia.

"Measures taken to prevent" relate to ensuring the safe operation fo the plant under normal operaating conditions or within specified process limits.  They are those measures which are intended to prevent the initiation of a sequence of events which could lead to a significant hazardous outcome and would include consideration of design, engineering standards, constructional and quality assurance, inspection and maintenance, and control systems insofar as these were concerned with controlling the process during normal operation.  These preventive measures derive their validity from the way in which the overall management of the plant and company exericises control over the relevant parts of the management system, expanding on the general information given in Section 4.1.3 (c) will be needed to support any arguments about the probability of initiating a sequence of events with a hazardous outcome.  In relation to the ammonia sphere example, "measures taken to prevent", would include a discussion of the arrangements for checking ammonia purity and special vessel inspections in connection with the problem of stress corrosion cracking.

"Measures taken to control" relate to the interventions permitted by the design of the plant (e.g., valves) or safety hardware (e.g., dump tanks) which may counter an event so that the dangerous substance is retained within the plant.  The operation of a "measure taken to control" assumes that a sequence of events has been initiated and the control is intended to prevent the sequence proceeding to a major accident.  In relation to the LPG sphere example, a spillage of LPG near the tank which ignited, could be prevented from escalating to a BLEVE (Boiling Liquid Expanding Vapour Explosion) of the whole sphere by the effective operation of a water spray system on the sphere.  The discussion of this control measure should include reference to the capacity of the water storage system, the water application rate; how the company ensures that the system will in fact operate effectively when required to do so; and the system reliability.

The second part of the response to this item covers "measures to minimise the consequence", which relate to those measures which can be taken after the major accident has actually occurred.  Examples of minimisation include:  bunding, water curtains, foam blankets, and emergency procedures.  In relation to the ammonia sphere, the effects of a release of ammonia from a pipe at a sufficient rate to lead to a lethal concentration of gas at (say) the nearest domestic dwellings might be minimised by the application of a water curtain around the sphere, and, through the emergency services, by evacuation of those people who are downwind of the escape.

*Background Information Required*                                                                   *159*

Section 4.1.3 (d) (iii) refers to the prevention or control of major accidents identified in Section 4.1.3 (d) (i). It is therefore not necessary to include information about measures directed solely at preventing small releases or those which have only minor consequences, unless these have the potential to escalate to a major accident.

Section 4.1.3 (d) (iii) by concentrating on preventive measures, is intended to draw out discussion on the positive aspects of the major hazard assessment. However, because the sequence of cause, effect and consequence are closely intertwined, it is not possible in practice to confine such a discussion to prevention without mentioning the potential consequences of the potential major accidents discussed in the major hazard assessment. Estimates of consequences will in any case be required in order to formulate adequate advice to the authority responsible for drawing up off-site emergency plans (see Section 5).

## (iv)    Emergency Procedures

Section 5 requires the developer to produce a plan for dealing with emergencies on-site. The operator/developer should be able to present the whole document for examination if necessary, but that is not the intention of this part of the major hazard assessment. On the other hand, it will not be sufficient merely to state that the emergency plan exists. The response to this item should describe the procedures in outline; indicate the nature and extent of emergencies with which the plans are intended to cope, drawing as necessary on the information in other parts of the major hazard assessment; mention those arrangements which may be critical to the success of the plans such as access for emergency services, the provision of adequate supplies of fire-fighting water, the remote siting of emergency control points and the evacuation of non-essential site personnel; and confirm that the plans have been discussed with the relevant outside bodies and practised with them.

## (v)    Meteorological Conditions

Data should be obtained from the nearest Meteorological Office weather station as to the prevailing weather conditions in the vicinity of the site, and if necessary, confirmed by actual measures on the site.

## (vi)    Numbers at Risk

This estimate should include those people normally working on the plant concerned and any major concentrations of people in the immediate vicinity of the plant, e.g., office buildings. As for Section 4.1.3 (b) (iv), exact numbers are not necessary.

# APPENDIX V

# Summary of Reactive Hazards

| Process Unit | Process Step | Nature of Hazard | Means of Control |
|---|---|---|---|
| 1. | Thin film evaporator. | Runaway polymerisation of Chemical A at $> 170^{\circ}C$ in the presence of traces of catalyst. | - Use of thin film evaporation. Steam at $150^{\circ}C$, water dousing system, relief to blowdown drum Temp. Alarms and interlocks.<br><br>- Small quantities only in process. |
| 2. | Reaction stepnitration of Chemical B. | Highly exothermic reaction with catalyst. | - Continuous reaction system with small quantities of reactions in large quantity of liquid catalyst recycle. |
| 3. | Reaction step-nitration of Chemical C. | Highly exothermic - reaction with $H_2SO_4$ as catalyst. | - Reactants metered by dosing pump system with special safeguards. Reaction takes place in pump after static mixer. Temp. alarms and interlocks to prevent runaway reactions.<br><br>- Small quantities of reactants. |
| 4. | Reboilers of distillation columns in purification stages. | Runaway polymerisation reaction. | - Small quantity of material with temperature control. Safe discharge of reboiler contents to an enclosed area away from other process equipment and personnel via a rupture disc relief system and vent angled at $45^{\circ}$ to horizontal $< 100$ kg products. |

| | | | |
|---|---|---|---|
| 4. | Reboilers of distillation columns in purification stages. | Runaway polymerisation reaction. | - Small quantity of material with temperature control. Safe discharge of reboiler contents to an enclosed area away from other process equipment and personnel via a rupture disc relief system and vent angled at $45^\circ$ to horizontal < 100 kg products. |
| 5. | Reaction step - Hydrolysis of Chemical D at $160^\circ$C and 4 atmos.<br><br>All reactants are added at start of batch and heated to operating temperature of $150^\circ$C. | Highly exothermic in 10 m3 reactors. Scale up from 100 litre pilot plant and operating experience on 3 $m^3$ reactor with similar reactants. | - Over designed external coil cooling system for heat of reaction of 41 kcal/mol cf. actual value of 14 kcal/mol.<br><br>- Manual blow-down on reactor.<br><br>- Semi-automatic blow-down system from control room.<br><br>- Computer control to minimise human error.<br><br>- Relief valve system to blow-down. |
| 6. | Two step reaction process. First step involving catalytic reduction of Chemical E to Chemical F is the hazardous step. | If insufficient active catalyst (i.e., due to poisoning or inadequate addition), and too low a reaction temperature, a build up of an intermediate product occurs. This intermediate can cause a runaway reaction if the temperature is raised too quickly. | - Extended batch reaction cycle from 1/2 hour to 3 hours.<br><br>- Keep reaction temperature > $80^\circ$C.<br><br>- Control flow of $H_2$.<br><br>- Shut off flow of $H_2$ if temperature < $80^\circ$C. |

| 7. | Catalytical reduction of Chemical G in a continuous reactor at 60 atmosphere's pressure. | Similar reaction hazard as Unit 6, if the intermediate forms due to insufficient active catalyst and too low temperature. This is the first time this reactor system has been used to produce Chemical H. Has been used for similar reactions and an extensive computer control/safety system has been developed. | - Fully computer controlled.<br><br>- Hydrogen addition regulated according to heat balance calculations.<br><br>- Auto shut-off of $H^2$ if heat balance shows generation of inter-mediate.<br><br>- Initially there will be no recycle of catalyst but may recycle as gain experience.<br><br>- Reactor blow-down system installed.<br><br>- Safety interlocks.<br><br>- Significant volume of chemicals present. |

# APPENDIX C

## Summary of Potential Sources of Ignition

## REDUCTION OF IGNITION RISK

Industrial plants contain a great number of possible ignition sources.It is common practice to reduce as far as practical possible ignition sources in areas where there is a risk of flammable releases. This is achieved by enforcing regulations that:
  a)  restrict activity in the area (using a "permit to work system")
  b)  only allow suitable equipment in the area
  c)  restrict access to the area (by fencing off the area)

## LIST OF IGNITION SOURCES

Lees ( 1980 ) lists likely ignition sources. The list is given here with suggestions on how ignition sources should be controlled within a restricted area.

| Ignition Source | Method of Control |
|---|---|
| 1.  Burner flames. | Layout of plant, trip systems. |
| 2.  Burning operations. | Permit-to-work. |
| 3.  Hot soot. | |
| 4.  Cigarettes. | No-smoking in restricted area. |
| 5.  Smouldering material. | |
| 6.  Hot process equipment. | Design. |
| 7.  Distress machinery. | Design and maintenance. |
| 8.  Small process fires. | |
| 9.  Weldings and cutting. | Permit-to-work. |
| 10. Mechanical sparks. | Special non-sparking tools are available. |
| 11. Vehicles. | Ordinary vehicles excluded from areas of flammable hazard. |
| 12. Arson. | Security procedures implemented in hazardous areas. |
| 13. Self-heating. | This occurs through slow oxidation of a solid.A typical source is oily rags."Good Housekeeping" should prevent this. |

14. Static electricity

This complex problem cannot be properly treated here. The reader is referred to specialist texts such as "Electrostatics in the Petroleum Industry" by Klinkenberg and van der Minne (1958) .

15. Electrical equipment.

Electrical equipment to be positioned in these areas have to be built to special standards. In the UK and in many other countries the areas at risk are classified according to the severity of the risk with a different electrical equipment standard for each classification.

# APPENDIX D

## The Properties of Some Hazardous Materials

### D.1 FLAMMABLE PROPERTIES

The models presented in Section 4 require upper flammability limits (UFLs) and lower flammability limits (LFLs). Table D.1 below gives some data for common hazardous materials presented as volume per cent in air. If data are not available for a particular material it is possible to estimate a flammability limit by taking data for a similar material and applying the following:

$$LFL_A = \frac{M_B}{M_A} LFL_B$$

where $M_A$ and $M_B$ are the relevant molecular weights.

| Compound | Lower (%v/v) | Upper (%v/v) |
|---|---|---|
| | Limits of flammability | |
| Acetone | 2.6 | 13.0 |
| Acetylene | 2.5 | 100.0 |
| Ammonia | 15.0 | 28.0 |
| Amylene | 1.8 | 8.7 |
| Benzene | 1.4 | 8.0 |
| n-Butane | 1.8 | 8.4 |
| i-Butane | 1.8 | 8.4 |
| Butene-1 | 2.0 | 10.0 |
| Butene-2 | 1.7 | 9.7 |
| Cyclohexane | 1.3 | 7.8 |
| Decane | 0.8 | 5.4 |
| Ethane | 3.0 | 12.4 |
| Ethylene | 2.7 | 36.0 |
| Ethylene dichloride | 6.2 | 15.9 |
| Ethylene oxide | 3.0 | 100.0 |
| Heptane | 1.2 | 6.7 |
| Hexane | 1.4 | 7.4 |
| Hydrogen | 4.0 | 75.0 |
| Methane | 5.0 | 15.0 |
| n Pentane | 1.8 | 7.8 |
| Propane | 2.1 | 9.5 |
| Propylene | 2.4 | 11.0 |
| Toluene | 1.3 | 7.0 |
| Vinyl chloride | 4.0 | 22.0 |
| 2,2-Dimethylpropane | 1.3 | 7.5 |
| 2,3-Dimethylpentane | 1.1 | 6.8 |

TABLE D.1 : *Flammability Limits of Some Common Materials*

## D.2 TOXIC PROPERTIES

During the last 10 years, probit equations have been derived for estimating, from dose relationships, the probability of affecting a certain proportion of the exposed population. These have been based almost exclusively on animal test data, which are often very imprecise anyway, and can sequentially they are not very reliable. In particular they do not possess the accuracy which could be ascribed to the formula as used in calculations of consequential effect, and they only apply to acute effects of accidental exposures. In recent years, some of the original equations have been subject to much scrutiny and criticism, and this has resulted in alternative equations being proposed.

In Table D.2 below are listed the latest, or only, equations available for a variety of toxic gases, but it is important that the origin of each equation is correctly referenced so that the effect of using alternative equations can be quickly established. The difference in results can be exceedingly large in some instances.

Probit Equations take the form:

$$Pr = At + Bt \ln (Cnt_e)$$

where     Pr is the probability function (expressed in units of standard deviation, but with an offset of +5 to avoid the use of negative values)
      At, Bt, and n are constants, and
      C is concentration of pollutant to which exposure is made, and is ppm v/v
      $t_e$ is the duration of exposure to the pollutant, measured in minutes.

Note: if the unit of C and/or te are changed, e.g. to mgm/litre v/v, then the values of the constants will change. All units should therefore be quoted.

| Material | $A_t$ | $B_t$ | n | Reference |
|---|---|---|---|---|
| Chlorine | -5.3 | 0.5 | 2.75 | DCMR 1984 |
| Ammonia | -9.82 | 0.71 | 2.0 | DCMR 1984 |
| Acrolein | -9.93 | 2.05 | 1.0 | USCG 1977 |
| Carbon Tetrachloride | 0.54 | 1.01 | 0.5 | USCG 1977 |
| Hydrogen Chloride | -21.76 | 2.65 | 1.0 | USCG 1977 |
| Methyl Bromide | -19.92 | 5.16 | 1.0 | USCG 1977 |
| Phosgene | -19.27 | 3.69 | 1.0 | USCG 1977 |
| Hydrogen Fluoride (monomer) | -26.4 | 3.35 | 1.0 | USCG 1978 |

TABLE D.2: *Toxic Properties of Some Materials*

# Bibliography

AMERICAN PETROLEUM INTITUTE (1969) Guide for Pressure Relief and Depressuring Systems; New York: API RP 521

BRIGGS, G.A (1969) Plume Rise *U.S. Atomic Energy Commision, Div. Tech., AEC Critical Rev.* Oak Ridge, Tenn.

BRIGGS, G.A. (1976) The Lift-Off of a Buoyant Release at Ground Level.

BRISCOE, F. and SHAW, P. (1978) Evaporation from Spills of Hazardous Liquids on Land and Water. *SRD R100.*

COX, R.A. and CARPENTER, R.J.(1980) Further Development of a Dense Vapor Cloud Dispersion Model for Hazard Analysis. In: *Heavy Gas and Risk Analysis, edited by Sittartwig.*

CLANCEY, V.J. (1972) Diagnostic Features of Explosion Damage *Sixth Int. Mgt of Forensic Science*

COX, B.G. and SAVILLE, G. (eds) (1975) High Pressure Safety Codes; London, Imperial College: *High Technology Technology Association.*

CRANE (1981) Flow of Fluids through Valves, Fittings and Pipes - Metric Edition. *Technical*

CUDE, A.L. (1975) The Generation Spread and Decay of Flammable Vapour Clouds. *IChemE Course "Process Safety - Theory and Practice"*, Teeside Polytechnic, Middlesborough, 7-10 July 1975.

D.I.P.P.R. (1985) Data Compilation Tables of Properties of Pure Compounds, *A.I.CheE.Design Institute for Physical Property Data.* New York.

EMERSON, M.C (1986) A new 'unbounded' jet dispersion model. *5th International Symposium on Loss Prevention and Safety Promotion in the Process Industries,* Cannes, September 1986.

ENGLAND, W.G. et al. (1978) Atmospheric Dispersion of LNG Vapour Cloud using SIGMET, a 3-D Time-Dependent Hydrodynamic Computer Model. *Heat Transfer and Fluid Mechanics Institute*, Washington.

FAUSKE, H.K., (1965). The Discharge of Saturated Water through Pipes. *CEP Symp. Series 59:*

FINNEY, D.J., (1971) Probit Analysis; London,Cambridge Univ. Press.

KLINKENBERG, A. and VAN DER MINNE, J.L. (1958) Electrostatics in the Pertroleum Industry, Amsterdam, Elsevier.

LEES, F.P., (1980) Loss Prevention in the Process Industries, London, Butterworths.

McADAMS, W.H. (1954) Heat Transmission, 3rd Edition, New York, McGraw Hill.

MONIN, A.S. (1962) *J. Geographys.Res.,* 67, 3103.

MOORHOUSE, J. and PRITCHARD, M.J. (1982) Thermal Radiation Hazards from Large Pool Fires and Fireballs. *A Literature Review. The Assessment of Major Hazards Symposium,* Manchester.

NIOSH, National Institute for Occupational Safety and Health; Item 37, Washington D.C.: U.S. Govt. Printing Office

PASQUILL, F. (1961) The Estimation of the Dispersion of Windborne Meterials. *Met. Mag.,* 901063), 33.

SUTTON, O.G. (1953) Atmospheric Diffusion, London: Van Nostrand.

ROBERTS, A.F. (1982). "The Effect of Conditions Prior to Loss of Containment in Fireball Behaviour." *The Assessment of Major Hazards Symposium*, Manchester.

TOEGEPAST NATURWETESHAPPELIJK ONDERZOEK (TNO), (197). Methods for the Calculation of the Physical Effects of the Escape of Dangerous Material ("TNO Yellow Book).

ULDEN, A.P. van, (1974). "On the Spreading of a Heavy Gas Released near the Ground." *1st International Loss Prevention Symposium*, The Hague/Delft.

# DISTRIBUTORS OF WORLD BANK PUBLICATIONS

**ARGENTINA**
Carlos Hirsch, SRL
Galeria Guemes
Florida 165, 4th Floor-Ofc. 453/465
1333 Buenos Aires

**AUSTRALIA, PAPUA NEW GUINEA, FIJI, SOLOMON ISLANDS, VANUATU, AND WESTERN SAMOA**
Info-Line
Overseas Document Delivery
Box 506, GPO
Sydney, NSW 2001

**AUSTRIA**
Gerold and Co.
A-1011 Wien
Graben 31

**BAHRAIN**
MEMBR Information Services
P.O. Box 2750
Manama Town 317

**BANGLADESH**
Micro Industries Development Assistance Society (MIDAS)
G.P.O. Box 800
Dhaka

**BELGIUM**
Publications des Nations Unies
Av. du Roi 202
1060 Brussels

**BRAZIL**
Publicacoes Tecnicas Internacionais Ltda.
Rua Peixoto Gomide, 209
01409 Sao Paulo, SP

**CANADA**
Le Diffuseur
C.P. 85, 1501 Ampere Street
Boucherville, Quebec
J4B 5E6

**COLOMBIA**
Enlace Ltda.
Carrera 6 No. 51-21
Bogota D.E.

Apartado Aereo 4430
Cali, Valle

**COSTA RICA**
Libreria Trejos
Calle 11-13
Av. Fernandez Guell
San Jose

**COTE D'IVOIRE**
Entre d'Edition et de Diffusion Africaines (CEDA)
04 B.P. 541
Abidjan 04 Plateau

**CYPRUS**
MEMRB Information Services
P.O. Box 2098
Nicosia

**DENMARK**
SamfundsLitteratur
Rosenoerns Alle 11
DK-1970 Frederiksberg C.

**DOMINICAN REPUBLIC**
Editora Taller, C. por A.
Restauracion
Apdo. postal 2190
Santo Domingo

**EGYPT, ARAB REPUBLIC OF**
Al Ahram
Al Galaa Street
Cairo

**FINLAND**
Akateeminen Kirjakauppa
P.O. Box 128
SF-00101
Helsinki 10

**FRANCE**
World Bank Publications
66, avenue d'Iéna
75116 Paris

**GERMANY, FEDERAL REPUBLIC OF**
UNO-Verlag
Poppelsdorfer Alle 55
D-5300 Bonn 1

**GREECE**
KEME
24, Ippodamou Street
Athens-11635

**GUATEMALA**
Librerias Piedra Santa
Centro Cultural Piedra Santa
11 calle 6-50 zona 1
Guatemala City

**HONG KONG, MACAU**
Asia 2000 Ltd.
6 Fl., 146 Prince Edward Road, W,
Kowloon
Hong Kong

**HUNGARY**
Kultura
P.O. Box 139
1389 Budapest 62

**INDIA**
Allied Publishers Private Ltd.
751 Mount Road
Madras—600 002

15 J.N. Heredia Marg
Ballard Estate
Bombay—400 038

13/14 Asaf Ali Road
New Delhi—110 002

17 Chittaranjan Avenue
Calcutta—700 072

Jayadeva Hostel Building
5th Main Road Gandhinagar
Bangalore—560 009

3-5-1129 Kachiguda Cross Road
Hyderabad—500 027

Prarthana Flats, 2nd Floor
Near Thakore Baug, Navrangpura
Ahmedabad—380 009

Patiala House
16-A Ashok Marg
Lucknow—226 001

**INDONESIA**
Pt. Indira Limited
Jl. Sam Ratulangi 37
Jakarta Pusat
P.O. Box 181

**IRELAND**
TDC Publishers
12 North Frederick Street
Dublin 1

**ISRAEL**
The Jerusalem Post
The Jerusalem Post Building
P.O. Box 81
Romema Jerusalem 91000

**ITALY**
Licosa Commissionaria Sansoni SPA
Via Lamarmora 45
Casella Postale 552
50121 Florence

**JAPAN**
Eastern Book Service
37-3, Hongo 3-Chome, Bunkyo-ku 113
Tokyo

**JORDAN**
Jordan Center for Marketing Research
P.O. Box 3143
Jabal
Amman

**KENYA**
Africa Book Service (E.A.) Ltd.
P.O. Box 45245
Nairobi

**KOREA, REPUBLIC OF**
Pan Korea Book Corporation
P.O. Box 101, Kwangwhamun
Seoul

**KUWAIT**
MEMRB
P.O. Box 5465

**MALAYSIA**
University of Malaya Cooperative Bookshop Limited
P.O. Box 1127, Jalan Pantai Baru
Kuala Lumpur

**MEXICO**
INFOTEC
Apartado Postal 22-860
Col. PE/A Pobre
14060 Tlalpan, Mexico D.F.

**MOROCCO**
Societe d'Etudes Marketing Marocaine
2 Rue Moliere, Bd. d'Anfa
Casablanca

**THE NETHERLANDS**
InOr Publikaties
Noorderwal 38
7241 BL Lochem

**NEW ZEALAND**
Hills Library and Information Service
Private Bag
New Market
Auckland

**NIGERIA**
University Press Limited
Three Crowns Building Jericho
Private Mail Bag 5095
Ibadan

**NORWAY**
Tanum-Karl Johan, A.S.
P.O. Box 1177 Sentrum
Oslo 1

**OMAN**
MEMRB Information Services
P.O. Box 1613, Seeb Airport
Muscat

**PAKISTAN**
Mirza Book Agency
65, Shahrah-e-Quaid-e-Azam
P.O. Box No. 729
Lahore 3

**PERU**
Editorial Desarrollo SA
Apartado 3824
Lima

**THE PHILIPPINES**
National Book Store
701 Royal Avenue
Metro Manila

**POLAND**
ORPAN
Palac Kultury i Nauki
00-901 WARSZAWA

**PORTUGAL**
Liveria Portugal
Rua Do Carmo 70-74
1200 Lisbon

**SAUDI ARABIA, QATAR**
Jarir Book Store
P.O. Box 3196
Riyadh 11471

**SINGAPORE, TAIWAN, BURMA, BRUNEI**
Information Publications
Private, Ltd.
02-06 1st Fl., Pei-Fu Industrial
Bldg., 24 New Industrial Road
Singapore

**SOUTH AFRICA**
*For single titles:*
Oxford University Press Southern Africa
P.O. Box 1141
Cape Town 8000

*For subscription orders:*
International Subscription Service
P.O. Box 41095
Craighall
Johannesburg 2024

**SPAIN**
Mundi-Prensa Libros, S.A.
Castello 37
28001 Madrid

**SRI LANKA AND THE MALDIVES**
Lake House Bookshop
P.O. Box 244
100, Sir Chittampalam A. Gardiner Mawatha
Colombo 2

**SWEDEN**
*For single titles:*
ABCE Fritzes Kungl. Hovbokhandel
Regeringsgatan 12, Box 16356
S-103 27 Stockholm

*For subscription orders:*
Wennergren-Williams AB
Box 30004
S-104 25 Stockholm

**SWITZERLAND**
Librairie Payot
6 Rue Grenus
Case postal 381
CH 1211 Geneva 11

**TANZANIA**
Oxford University Press
P.O. Box 5299
Dar es Salaam

**THAILAND**
Central Department Store
306 Silom Road
Bangkok

**TRINIDAD & TOBAGO, ANTIGUA, BARBUDA, BARBADOS, DOMINICA, GRENADA, GUYANA, JAMAICA, MONTSERRAT, ST. KITTS AND NEVIS, ST. LUCIA, ST. VINCENT & GRENADINES**
Systematics Studies Unit
55 Eastern Main Road
Curepe
Trinidad, West Indies

**TURKEY**
Haset Kitapevi A.S.
469, Istiklal Caddesi
Beyoglu-Istanbul

**UGANDA**
Uganda Bookshop
P.O. Box 7145
Kampala

**UNITED ARAB EMIRATES**
MEMBR Gulf Co.
P.O. Box 6097
Sharjah

**UNITED KINGDOM**
Microinfo Ltd.
P.O. Box 3
Alton, Hampshire GU 34 2PG
England

**VENEZUELA**
Libreria del Este
Aptdo. 60.337
Caracas 1060-A

**YUGOSLAVIA**
Jugoslovenska Knjiga
YU-11000 Belgrade Trg Republike

**ZIMBABWE**
Textbook Sales Pvt. Ltd.
Box 3799
Harare